The Scientific Papers of Grant Allen

Volume I

Charles Grant Blairfindie Allen was born on February 24th, 1848 at Alwington, near Kingston, Canada West (now part of Ontario).

Home schooled until 13 when his family moved to England, Grant was to become a highly regarded science writer who branched out to a fiction career and became enormously popular.

His work helped propel several genres of fiction and whilst his career was short it was enormously productive.

Grant's scientific background enabled him to root much of his work in a plausibility that was denied to others. He had little fear in challenging a society that treated women as second class citizens and creating best sellers from such works.

On October 25th 1899 Grant Allen died at his home in Hindhead, Haslemere, Surrey, England. He died just before finishing Hilda Wade. The novel's final episode, which he dictated to his friend, doctor and neighbour Sir Arthur Conan Doyle from his bed appeared under the appropriate title, The Episode of the Dead Man Who Spoke in 1900.

Index of Contents

THE ORIGIN OF FRUITS

In the whole museum of Nature the eye of the artist can find nothing lovelier than flowers; but the second rank in beauty may be fairly claimed on behalf of fruits. Whether we look at the golden oranges, the pink-cheeked mangoes, the purple star-apples, and the scarlet capsicums of the south, or at our own crimson cherries, blushing grapes, bright holly-berries, and rosy apples, we are equally struck with the delicacy of their melting tints and the graceful curves of their rounded form. Our painters have reveled in their rich coloring; and even our sculptors, whose fastidious art compels them to reject that meretricious charm, have loved to chisel their swelling contours in snowy stone. As they hang pendent from their native boughs, clustering in brilliant masses, or scattered here and there as points of brighter light amid the dark foliage which throws up in strong relief their exquisite hues, we may recognize in their beauty the ultimate source of all that refined pleasure which mankind derives from the varied shades of earth and sea and sky, of flower and bird and butterfly, and even of the "human face divine" itself. From the contemplation of ruddy or snowy berries in primeval forests the frugivorous ancestors of our race first acquired the taste for brilliant hues, whose final outcome has produced at length our modern picture-galleries and palaces, our flower-gardens and conservatories, our household ornament and our decorative art.

In a previous paper on "The Origin of Flowers,"[1] we endeavored to trace the mutual reactions of insects and blossoms upon one another's forms and hues. But we then deferred for a while the consideration of the further question—"Why do human beings admire these bright whorls of colored leaves, whose primitive function consisted in the attraction of bees and butterflies? Through what community of origin or nature does the eye of man find itself agreeably stimulated by the tints which were first developed to suit the myriad facets of primeval insects?" The answer to this question we have now to attempt, by showing the various steps through which the coverings of certain seeds acquired, for the vertebrate orders—the birds and quadrupeds—exactly the same allurements of color, scent, and taste, which flowers had already acquired for the articulate orders—the bees and butterflies. To the attractive hues of fruit, I believe, we must ultimately trace back our whole artistic pleasure in the pure physical stimulation of beautiful colors, displayed by natural objects or artificial products.

Our present inquiry, then, will yield us some account of that primitive delight in red, purple, orange, and yellow, which we usually take for granted as an innate instinct of humanity, savage or civilized. When, some few months back, we analyzed the various elements of pleasure which make up our aesthetic enjoyment of a daisy, we were compelled, for the time being, to leave the original beauty of its pink-and-white rays wholly unexplained. We regarded the delight in color, relatively to the subject we were then examining, as an ultimate and indecomposable factor in our developed consciousness. To-day, however, I hope we shall be able to go a little further back, and to show that this delight, like all other feelings of our nature, is no mere chance and meaningless accident, but the slow result of a long adaptation whereby man has gradually become fitted to the high and responsible station which he now occupies at the head of organic existence.

The sole object of flowering is the production of seeds—that is to say, of embryo plants, destined to replace their parents, and continue the life of their species to future generations. Flowers and seeds go together; every flower producing seed, and every seed springing from a flower. Ferns and other like plants, which have no blossoms, bring forth spores which grow into shapeless little fronds, instead of true seeds containing a young plantlet. But all flowering species produce some kind of genuine fruit, supplied with more or less nutriment for the tender embryo in its earlier days. And this matter of nutriment is so important to a right comprehension of our subject that I venture, even at the imminent peril of appearing dull, to digress a little into the terrible mysteries of Energy, which comprise the whole difficulty of the question.

Wherever movement is taking place in any terrestrial object, the energy which moves it has been directly or indirectly supplied from the sun. In the green parts of plants, the solar rays are perpetually producing a separation of carbon and oxygen, the former element being stored up in the tissues themselves, while the latter is turned loose upon the atmosphere in a free state. Whenever they recombine, motion and heat will result, as we see alike in our grates, our steam-engines, and our own bodies. An animal is a sort of machine—viewed from a purely physical standpoint—in which the energetic materials laid up by plants are being reconverted into the warmth which reveals itself to our touch, and the evident movement which we see in its limbs. The vegetable or animal substances which are capable of yielding these energies to our bodies we know as food or nutriment. They perform exactly the same part in the physical economy of men or beasts as that which fuel performs in the physical economy of the steam-engine. Of course, from the mental point of view, we have the immense difference between a self-conscious, self-guiding organism, and a dead machine requiring to be supplied and regulated by an external consciousness; yet in the fundamental physical necessity for energetic material, either as food or as fuel, both mechanisms follow essentially the self-same mechanical laws.

But what has all this to do with the origin of fruits? Very little at first sight, indeed, yet everything when we look at the bottom of the question. In fact, what is thus true of animals and steam-engines is equally true of plants. No motion can take place in a growing shoot without the aid of solar energy, directly supplied by the sunshine, or indirectly laid by in the older tissues. In the green parts of a plant this energy is immediately derived from the bounteous light which bathes and vivifies the leaves on every side; but in many other portions of the vegetable organism, energies previously accumulated by older organs are perpetually being utilized, for the production of movement and growth, by lazy structures which cannot work for themselves, and so feed upon the useful materials collected for them by more industrious members of the plant-commonwealth. Especially is this the case with those expensive organs which are concerned in perpetuating the species to future generations. A flower or a seed cannot directly transform waves of light into chemical separation of atoms; they depend for their growth and the due performance of their important functions upon similar separations already carried on for their behoof by the green leaves on whose bounty they rely for proper subsistence. Carbon, set free from oxygen in the leaves, has been carried to them in loose combinations by the sap; and as the bud unfolds or the seed germinates, the oxygen once more unites with this carbon (just as it unites in the furnace of the steam-engine, or the recesses of the animal body), and motion is thereby rendered possible. But without such an access of free oxygen to recombine with the energetic materials, the blossom or the embryo could never grow at all. So we may regard these portions of a plant, incapable of self-support, and dependent for their due function upon energetic compounds laid by elsewhere, as the exact analogues of the animal or the steam-engine. They are in fact similar mechanisms, where food is being used up, and fuel is being consumed; and we find accordingly, as we might naturally expect, not only that motion results, but also that heat is evolved in quantities quite sufficient to be measured by very delicate thermometers.

Now, every growing portion of a plant shares, more or less, in this animal function of feeding upon previously-fabricated nutriment. But there are two sets of organs, both intended ultimately to subserve the same purpose, in which that function becomes especially apparent. The first is in the case of the whole regular reproductive mechanism, including in that term buds, flowers, fruits, and seeds; the second is in the case of such subsidiary reproductive devices as tubers, rhizomes, corms, and all the other varieties of underground stems or roots, which botanists divide into so many puzzling technical classes, while ordinary people are content to lump them roughly together as bulbs. If we glance briefly at each of these two cases, we shall be able to comprehend more fully their connection with the doctrine of energy, and also to see more clearly the problem before us when we endeavor to unravel the origin of fruits.

A germinating pea or a young blade of wheat is supplied by its parent with a large stock of nutriment in the shape of starch, albumen, or other common food-stuffs. If we were to burn the wheat instead of planting it, the energy contained in its substance would be given off during the act of combustion as light and heat. If, again, we were to adopt a more usual course, by grinding, baking, and eating it, then the inclosed energy would minister to the warmth of our bodies, and do its little part in enabling us to walk a mile or to lift a heavy weight. But if, in lieu of either plan, we follow the original design of Nature by covering the seed with moist earth, the chemical changes which take place within it, still resulting in heat and motion, produce that special form of movement which we know as germination. New cells form themselves about the feathery head, a little sprout pushes timidly its way through the surrounding soil, and soon a pair of rounded leaves or a spike of pointed blades may be seen spreading a mass of delicate green toward the open sunlight overhead. By the time that all the stored-up nutriment contained in the seed has been thus devoured by the young plantlet, these green surfaces are in a position to assimilate fresh material for themselves, from the air which bathes them on every side, under the energetic influence of the sunbeams that fall each moment on their growing cells. But I need hardly point out the exact analogy which we thus perceive between the earliest action of the young plant and the similar actions of the frugivorous animals which subsist upon the food intended for its use.

If, however, we look at the second great case, that of bulbs and tubers, we shall see the same truth still more clearly displayed. You cannot grow a blade of wheat or a sprouting pea in the dark. The seed will germinate, it is true; but, as soon as the primitive store of nutriment has been used up, it will wither away and die. Naturally enough, when all its original energy is gone, and no new energy is afforded to it from without in the form of sunshine, it cannot miraculously make growth for itself out of nothing. But if you put a hyacinth bulb in a dark cellar, and supply it with a sufficiency of water, it will grow and blossom almost as luxuriantly as in a sunny window. Now, what is the difference between these two cases? Simply this: the wheat-grain or the pea has only nutriment enough supplied it by the parent-plant to carry it over the first few days of its life, until it can shift for itself; while the hyacinth has energetic materials stored up in its capacious bulb to keep it in plenty during all the days of its summer existence. If we plant it in an open spot where it can bask in the bright sunshine, it will produce healthy green leaves, which help it to flower and to carry on its other physiological actions without depending entirely upon its previous accumulations; but if we place it in some dark corner, away from the sun, though its leaves will be blanched and sickly-looking, it will still have sufficient nutriment of its own to support it through the blossoming season without the external aid of fresh sunshine.

Where did this nutriment come from, however? It was stored up, in the case of the seed, by the mother-plant; in the case of the bulb, by the hyacinth itself. The materials produced in the leaves were transferred by the sap into the flower or the stem, and were there laid by in safety till a need arose for their expenditure. All last year—perhaps for many years before—the hyacinth-leaves were busily engaged in assimilating nutritive matter from the air about them, none of which the plant was then permitted to employ in the production of a blossom, but all was prudently treasured up by the gardener's care in the swelling bulb. This year, enough nourishment has been laid by to meet the cost of flowering, and so our hyacinth is enabled to produce, through its own resources, without further aid from the sun, its magnificent head of bright-colored and heavily-scented purple bells.

Each species of plant must, of course, solve for itself the problem, during the course of its development, whether its energies will be best employed by hoarding nutriment for its own future use in bulbs and tubers, or by producing richly-endowed seeds which will give its offspring a better chance of rooting themselves comfortably, and so surviving in safety amid the ceaseless competition of rival species. The various cereals, such as wheat, barley, rye, and oats, have found it most convenient to grow afresh with each season, and to supply their embryos with an abundant store of food for their sustenance during the infant stage of plant-life. Their example has been followed by peas and other pulses, by the wide class of nuts, and by the majority of garden-fruits. On the other hand, the onion and the tiger-lily store nutriment for themselves in the underground stem, surrounded by a mass of overlapping or closely-wound leaves, which we call a bulb; the iris and the crocus lay by their stock of food in a woody or fleshy stalk; the potato makes a rich deposit of starch in its subterraneous branches or tubers; the turnip, carrot, radish, and beet, use their root as the storehouse for their hoarded food-stuffs; while the orchis produces each year a new tubercle by the side of its existing root, and this second tubercle becomes in turn the parent of the next year's flowering stem. Perhaps, however, the common colchicum or meadow-saffron affords the most instructive instance of all; for during the summer it sends up green leaves alone, which devote their entire time to the accumulation of food-stuffs in a corm at their side; and, when the autumn comes round, this corm produces, not leaves, but a naked flower-stalk, which pushes its way through the moist earth, and stands solitary before the October winds, depending wholly upon the stock of nutriment laid up for it in the corm.

If we look at the parts of plants which are used as food by man of other mammals, we shall see even more clearly the community of nature between the animal functions and those of seeds, flowers, and bulbs. It is true that the graminivorous animals, like deer, sheep, cows, and horses, live mainly off the green leaves of grasses and creeping plants. But we know how small an amount of food they manage to extract from these fibrous masses, and how constantly their whole existence is devoted to the monotonous and imperative task of grazing for very life. Those animals, however, who have learned to live at the least cost to themselves always choose the portions of a plant which it has stored with nourishment for itself or its offspring. Men and monkeys feed naturally off fruits, seeds, and bulbs. Wheat, maize, rye, barley, oats, rice, millet, peas, vetches, and other grains or pulses, form the staple sustenance of half mankind. Other fruits largely employed for food are plantains, bananas, bread-fruit, dates, cocoanuts, chestnuts, mangoes, mangostines, and papaws. Among roots, tubers, and bulbs, stored with edible materials, may be mentioned beets, carrots, radishes, turnips, swedes, ginger, potatoes, yam, cassava, onions, and Jerusalem artichokes. But if we look at the other vegetables used as food, we shall observe at once that they are few in number, and unimportant in economical value. In cabbages, Brussels sprouts, lettuce, succory, spinach, and water-cress, we eat the green leaves; yet nobody would ever dream of making a meal off any of these poor food-stuffs. The stalk or young sprout forms the culinary portion of asparagus, celery, seakale, rhubarb, and angelica, none of which vegetables are remarkable for their nutritious properties. In all the remaining food-plants, some part of the flowering apparatus supplies the table, as in true artichokes, where we eat the receptacle, richly stocked with nutriment for the opening florets; or in cauliflower, where we choose the young flower-buds themselves. In short, we find that men and the higher animals generally support themselves upon those parts of plants in which energy has been accumulated either for the future growth and unfolding of the plant itself, or for the sustenance of its tender offspring.

And now, after this long preamble, let us come back to our original question, and seek to discover what is the origin of fruits.

In botanical language, every structure which contains the seeds resulting from the fertilization of a single blossom is known as a fruit, however hard, dry, and unattractive, may be its texture or appearance. But I propose at present to restrict the term to its ordinary meaning in the mouths of every-day speakers, and to understand by it some kind of succulent seed-covering, capable of being used as food by man or other vertebrates. And our present object must be simply to discover how these particular coverings came to be developed in the slow course of organic evolution,

Doubtless the earliest seeds differed but little from the spores of ferns and other flowerless plants in the amount of nutriment with which they were provided and the mode in which they were dropped upon the nursing soil beneath. But as time went on, during the great secondary and tertiary ages of geology, throughout whose long course first the conifers and then the true flowering plants slowly superseded the gigantic horsetails and tree-ferns of the coal-measures, many new devices for the dispersion and nutrition of seeds were gradually developed by the pressure of natural selection.[2] Those plants which merely cast their naked embryos adrift upon the world to shift for themselves in the fierce struggle of stout and hardy competitors must necessarily waste their energies in the production of an immense number of seeds. In fact, calculations have been made which show that a single scarlet corn-poppy produces in one year no less than 50,000 embryos; and some other species actually exceed this enormous figure. If, then, any plant happens by a favorable combination of circumstances to modify the shape of its seed in such a manner that it can be more readily conveyed to open or unoccupied spots, it will be able in future to economize its strength, and thus to give both itself and its offspring a better chance in the struggle for life. There are many ways in which natural selection has effected this desirable consummation.

The thistle, the dandelion, and the cotton-bush, provide their seeds with long tufts of light hair, thin and airy as gossamer, by which they are carried on the wings of the wind to bare spaces, away from the shadow of their mother-plant, where they may root themselves successfully in the vacant soil. The maple, the ash, and the pine, supply their embryos with flattened wings, which serve them in like manner not less effectually. Both these we may classify as wind-dispersed seeds. A second set of plants have seed-vessels which burst open explosively when ripe, and scatter their contents to a considerable distance. The balsam forms the commonest example in our European gardens; but a well-known tropical tree, the sand-box, displays the same peculiarity in a form which is almost alarming, as its large, hard, dry capsules fly apart with the report of a small pistol, and drive out the disk-shaped nuts within so forcibly as to make a blow on the cheek decidedly unpleasant. These we may designate as self-dispersed seeds. Yet a third class may be conveniently described as animal-dispersed, divisible once more into two sub-classes, the involuntarily and the voluntarily aided. Of the former kind we have examples in those seeds which, like burs and cleavers, are covered with little hooks, by whose assistance they attach themselves to the fur or wool of passers-by. The latter or voluntarily aided sort are exemplified in fruits proper, the subject of our present investigation, such as apples, plums, peaches, cherries, haws, and bramble-berries. Every one of these plants is provided with hard and indigestible seeds, coated or surrounded by a soft, sweet, pulpy, perfumed, bright-colored, and nutritious covering, known as fruit. By all these means the plant allures birds or mammals to swallow and disperse its undigested seed, giving in, as it were, the pulpy covering as a reward to the animal for the service thus conferred. But before we go on to inquire into the mode of their development we must glance aside briefly at a second important difference in the constitution of seeds.

If we plant a grain of mustard-seed in moist earth and allow it to germinate, we shall see that its young leaves begin from the very first to grow green and assimilate energetic matter from the air around them. They are, indeed, compelled to do so, because they have no large store of nutriment laid up in the seed-leaves for their future use by the mother-plant. But if we treat a pea in the same manner, we shall find that it long continues to derive nourishment from the abundant stock of food treasured up in its big, round seed-leaves. Now of course any plant which thus learns to lay by in time for the wants of its offspring gives its embryo a far better chance of surviving and leaving descendants in its turn, than one which abandons its infant plants to their own unaided resources in a stern battle with the unkindly world. Exactly the same difference exists between the two cases as that which exists between the wealthy merchant's son, launched on life with abundant capital accumulated by his father, and the street Arab, turned adrift, as soon as he can walk alone, to shift or starve for himself in the lanes and alleys of a great city.

So, then, as plants went on varying and improving under the stress of over-population, it would naturally result that many species must hit independently upon this device of laying by granaries of nutriment for the use of their descendants. But side by side with the advancing development of vegetable life, animal life was also developing in complexity and perfect adaptation to its circumstances. And herein lay a difficult dilemma for the unhappy plant. On the one hand, in order to compete with its neighbors, it must lay up stores of starch and oil and albumen for the good of its embryos; while, on the other hand, the more industriously it accumulated these expensive substances, the more temptingly did it lay itself open to the depredations of the squirrels, mice, bats, monkeys, and other clever thieves, whose number was daily increasing in the forests round about. The plant becomes, in short, like a merchant in a land exposed to the inroads of powerful robbers. If he does not keep up his shop with its tempting display of wares, he must die for want of custom; if he shows them too readily and unguardedly, he will lay himself open to be plundered of his whole stock in-trade. In such a case, the plant and the merchant have recourse to the self-same devices. Sometimes they surround themselves with means of defense against the depredators; sometimes they buy themselves off by sacrificing a portion of their wealth to secure the safety of the remainder. Those seeds which adopt the former plan we call nuts, while to those which depend upon the latter means of security we give the name of fruits.

A nut is a hard-coated seed, which deliberately lays itself out to escape the notice and baffle the efforts of monkeys and other frugivorous animals. Instead of bidding for attention by its bright hues, like the flower and fruit, the nut is purposely clad in a quiet coat of uniform green, indistinguishable from the surrounding leaves, during its earlier existence; while afterward it assumes a dull-brown color as it lies upon the dusky soil beneath. Nuts are rich in oils and other useful food-stuffs; but to eat these is destructive to the life of the embryo; and therefore the nut commonly surrounds itself with a hard and stony shell, which defies the stoutest teeth to pierce its thickened walls. Outside this solid coating it often spreads a softer covering with a nauseous, bitter taste, so familiar to us all in the walnut, which at once warns off the enemy from attacking the unsavory morsel. Not content with all these protective devices, of color, taste, and hardness, the nut in many cases contains poisonous juices, and is thickly clad in hooked and pointed mail, which wounds the hands or lips of the would-be robber. In brief, a nut is a seed which has survived in the struggle for life by means of multiplied protections against the attacks of enemies. We cannot have a better instance of these precautions than the common cocoanut palm. Its seed hangs at a great height from the ground, on a tall and slender stem, unprovided with branches which might aid the climber, and almost inaccessible to any animal except the persevering monkey. Its shell is very thick and hard, so extremely impermeable that a small passage has to be left by which the germinating shoot may push its way out of the stronghold where it is born. Outside this shell, again, lies a thick matting of hairy fibres, whose elasticity breaks its fall from the giddy height at which it hangs. Yet, in spite of all these cunning precautions, even the cocoanut is not quite safe from the depredations of monkeys, or, stranger still, of tree-climbing crabs. The common Brazil nuts of our fruiterers' shops are almost equally interesting, their queer, shapeless forms being closely packed together, as they hang from their native boughs, in a hard outer shell, not unlike that of the cocoanut. It must be very annoying to the unsuspecting monkey, who has succeeded after violent efforts in breaking the external coat, to find that he must still deal with a mass of hard, angular, and uncanny nuts, which sadly cut his tender gums and threaten the stability of his precious teeth—those invaluable tools which serve him well in the place of knives, hammers, scissors, and all other human implements.

A fruit, on the other hand, lays itself open in every way to attract the attention of animals, and so to be dispersed by their aid, often amid the nourishing refuse of their meals. It is true that, with the fruit as with the nut, to digest the actual seed itself would be fatal to the life of the young plant. But fruits get over this difficulty by coating their seeds first with a hard, indigestible shell, and then with a soft, sweet, pulpy, and nutritious outer layer. The purely accidental or functional origin of this covering is testified by the immense variety of ways in which it has been developed. Sometimes a single seed has shown a slight tendency to succulence in its outer coat, and forthwith it has gone on laying up juices from generation to generation, until it has developed into a one-seeded berry. Sometimes a whole head of seeds has been surrounded by a fleshy stem, and the attention of animals has thenceforward encouraged its new habit by insuring the dispersion of its embryos. A few of the various methods by which fruits attain their object we shall examine in detail further on;

it will suffice for the present to point out that any property which secured for the seed dispersion by animal agency would at once give it an advantage over its fellows, and thus tend to be increased in all future generations.

So, then, as birds, squirrels, bats, monkeys, and the higher animals generally, increased on the face of the earth, every seed which showed a tendency to surround itself with succulent pulp would obviously gain a point thereby in its rivalry with other species. Accordingly, as we might naturally expect, fruits, which have been developed to suit the taste of birds and mammals, are of much more recent geological origin than flowers, which have been developed to suit the taste of insects. For example, there is no family of plants which contains a greater number of fruity seeds than the rose tribe, in which are comprised the apple, pear, plum, cherry, blackberry, raspberry, strawberry, quince, medlar, loquat, peach, apricot, and nectarine, besides the humbler hips, haws, sloes, and common hedge-fruits, which, though despised by lordly man, form the chief winter sustenance of such among our British birds as do not migrate to warmer climates during our chilly December days. Now, no trace of the rose tribe can be discovered until late in tertiary times; in other words, no fruit-bearers appear before the evolution of the fruit-eaters who called them into existence: while, on the other hand, the rapid development and variation of the tribe in the succeeding epoch show how great an advantage it derived from its tendency to produce edible seed-coverings.

But not only must these coverings be succulent and nutritious, they must also be conspicuous and alluring. For the attainment of these objects the fruit has recourse to just the same devices which had already been so successfully initiated by the insect-fertilized flowers. It collects into its pulpy substance a quantity of that commonly-diffused vegetable principle which we call sugar. Now sugar, from its crystalline composition, is peculiarly adapted for acting upon the exposed nerves of taste in the tongue of vertebrates; and the stimulation which it affords, like all healthy and normal ones, when not excessive in amount, is naturally pleasurable to the excited sense. Of course, in our own case, the long habituation of our frugivorous ancestors to this particular stimulant has rendered us peculiarly sensitive to its effects. But even from the first, there can be little doubt that a body so specially fitted to arouse sensation in the gustatory nerve must have afforded pleasure to the unspecialized palates of birds and rodents: for we know that even in the case of naturally carnivorous animals, like dogs, a taste for sugar is extremely noticeable. So, then, the sweet juices of the fruit were early added to its soft and nutritive pulp as an extra attraction for the animal senses.

Perfume, of course, follows in the wake of sweetness. Indeed, the difference between taste and smell is much smaller than most people imagine. When tiny floating particles of a body, in the gaseous state, affect certain exposed nerves in the cavity of the nose, we call the resulting sensation an odor; when larger particles of the same body, in the liquid or dissolved state, affect similar exposed nerves in the tongue, we call the resulting; sensation a taste. But the mechanism of the two senses is probably quite similar, while their exciting causes and their likes or dislikes are almost identical. As our great psychological teacher, Mr. Herbert Spencer, well puts it, "smell is anticipatory taste." So we need not be surprised to find that the delicate fragrance of peaches, strawberries, oranges, and pineapples, is a guide to their edibility, and a foreshadowing of their delicious flavor, leading us on by an instinctive action to place the savory morsels between our lips.

But the greatest need of all, if the plant would succeed in enticing the friendly parrot or the obsequious lemur to disperse its seed, is that of conspicuousness. Let the fruit be ever so luscious and ever so laden with sweet sirups, it can never secure the suffrages of the higher animals if it lies hidden beneath a mass of green foliage, or clothes itself in the quiet garb of the retiring nut. To attract from a distance the eyes of wandering birds or mammals, it must dress itself up in a gorgeous livery of crimson, scarlet, and orange. The contrast between nuts and fruits is exactly parallel to the contrast between the wind fertilized and the insect-fertilized flowers. An apple-tree laden with its red-cheeked burden, an orange-bough weighed down with its golden spheres, a rowan or a holly-bush displaying ostentatiously its brilliant berries to the birds of the air, is a second edition of the roses, the rhododendrons, and the May-thorns, which spread their bright petals in the spring before the fascinated eyes of bees and butterflies. Some gay and striking tint, which may contrast strongly with the green foliage around, is needed by the developing fruit, or else its pulpiness, its sweetness, and its fragrance, will stand it in poor stead beside its bright-hued compeers.

How fruits began to acquire these brilliant tints is not difficult to see. We found already in the case of flowers that all external portions of a plant, except such green parts as are actually engaged in assimilating carbon, under the influence of solar energies, show a tendency to assume tints other than green. This tendency would, of course, be checked by natural selection in those seeds which, like nuts, are destroyed by animals, and so endeavor to escape their notice; while it would be increased by natural selection in those seeds which, like fruits proper, derive benefit from the observation of animals, and so endeavor to attract their attention. But it is noticeable that fruits themselves are sour, green, and hard, during their unripe stage—that is to say, before the seeds are ready to be severed from the mother plant; and that they only acquire their sweet taste, brilliant color, and soft pulp, just at the time when their mature seeds become capable of a separate existence.

Perhaps, however, the point which most clearly proves the purely functional origin of fruits is found in the immense variety of their structure, a variety far surpassing that of any other vegetable organ. It does not matter at all what portion of the seed-covering or its adjacent parts happens first to show the tendency toward succulence, sweetness, fragrance, and brilliancy. It serves the attractive purpose equally well whether it be calyx, or stalk, or skin, or receptacle. Just as, in the case of flowers, we found that the colored portion might equally well consist of stamens, petals, sepals, bracts, or spathe—so, but even more conspicuously, in the case of fruits, the attractive pulp may be formed of any organ whatsoever which exhibits the least tendency toward a pulpy habit, and an accumulation of saccharine deposits.

Thus, in the pomegranate, each separate seed is inclosed in a juicy testa or altered shell; in the nutmeg and the spindle-tree, an aril or purely gratuitous colored mass spreads gradually over the whole inner nut; in the plum and cherry, a single part, the pericarp, divides itself into two membranes, whereof the inner or protective coat is hard and stony, while the outer or attractive coat is soft, sweet, and bright colored; in the strawberry, the receptacle, which should naturally be a mere green bed for the various seed-vessels, grows high, round, pulpy, sweet, and ruddy; in the rose, the fruit-stem expands into a scarlet berry, containing the seed-vessels within, which also happens in a slightly different manner with the apple, pear, and quince; while in the fig, a similar stem incloses the innumerable seeds belonging to a whole colony of tiny blossoms, which thus form a compound fruit, just as the daisy-head, with its mass of clustered florets, forms a composite flower. Strangest of all, the common South American cashew-tree produces its nut (which is the true fruit) at the end of a swollen, pulpy, colored stalk, and so preserves its embryo by the vicarious sacrifice of a fallacious substitute. These are only a few out of the many ways in which the selective power of animals has varied the surroundings of different seeds to serve a single ultimate purpose.

Nor is any plan too extravagant for adoption by some aberrant species. What seed-organ could seem less adapted for the attraction of animals than a cone like that of pines and fir-trees? Yet even this hard, scaly covering has been modified, in the course of ages, so as to form a fruit. In the cypress, with its soft young cones, we can see dimly the first step in the process; in the juniper, the cone has become quite succulent and berry-like; and finally, in the red fruit of the yew, all resemblance to the original type is entirely overlaid by its acquired traits.

Equally significant is the fact that closely-allied species often choose totally different means for attracting or escaping observation. Thus, within the limits of the rose tribe itself we get such remarkable variations as the strawberry, where the receptacle forms the fruit; the apple, which depends on the peduncle, or swollen stalk, for its allurement; the raspberry, where each seed-vessel of the compound group has a juicy coating of its own, and so forth: while, on the other hand, the potentilla has no fruit at all, in the popular sense of the word; and the almond actually diverges so far from the ordinary habits of the tribe as to adopt the protective tactics of a nut. Similarly, in the palm tribe, while most species fortify themselves against monkeys by shells of extravagant hardness, as we see in the vegetable ivory, and the solid coquilla-nuts from which door-handles are manufactured, a few kinds, like the date and the doom-palm, trust rather to the softness and sweetness of their pulp, as aids to dispersion. The truth which we learn from these diverse cases may be shortly summed up thus: Whatever peculiarities tend to preserve the life of a species, in whatever opposite ways, equally aid it in the struggle for life, and may be indifferently produced in the most closely-related types.

And now let us glance for a moment—less fully than the subject demands, for this long exposition has run away with our space—at the reactive effects of fruit upon the animal eye. We took it for granted above that birds and mammals could discriminate between the red or yellow berry and the green foliage in whose midst it grows. Indeed, were other proof wanting, we should be justified in concluding that animals generally are possessed of a sense for the discrimination of color, from the mere fact that all those fruits and flowers which depend for their dispersion or fertilization upon animal agency are brightly tinted, while all those which depend upon the wind, or other insentient energies, are green or dull-brown in hue. But many actual observations, too numerous to be detailed here, also show us, beyond the possibility of error, that the higher animals do, as a matter of fact, possess a sense of color, differing in no important particular from that of civilized man.

Whether this sense was developed, however, by the constant search for berries and insects, or whether it was derived from a still earlier ancestry, it would be very difficult to decide. It is possible that, as we saw reason to believe in the case of the flowers and the insect vision, the colors of fruits and the color-sense of birds and mammals may have gone on developing side by side; each plant surviving in proportion as its seeds grew more and more distinctive, and each animal, in turn, standing a better chance of food in proportion as its discrimination of such seeds grew more and more acute. But as there are excellent reasons for crediting fishes and reptiles also with a high faculty for the perception of color, it may be safer to conclude that the sense was inherited by birds and mammals from our common vertebrate progenitors, being only quickened and intensified by the reactive influence of fruits.

Yet it must be remembered that the earliest fruit-eaters, though they might find the scarlet, crimson, or purple coats of their food an aid to discrimination in the primeval forests, would not necessarily derive any pleasure from the stimulation thus afforded That pleasure has been slowly begotten in all frugivorous races by the constant use of these particular nerves in the search for food, which has at last produced in them a calibre and a sensitiveness answering pleasurably to the appropriate stimulation. Just as the peach, which a dog would reject, has become delicious to our sense of taste; just as the pineapple, at which he would sniff unconcernedly, has become exquisite to our sense of smell—so the pure tints of the plum, the orange, the mango, and the pomegranate, which he would disregard, have become lovely to our sense of color. And, further still, just as we transfer the tastes formed in the first two cases to the sweetmeats of the East, or to the violets, hyacinths, and heliotropes of our gardens, so do we transfer the taste formed in the third case to our gorgeous peonies, roses, dahlias, crocuses, tiger-lilies, and chrysanthemums; to our silks, satins, damasks, and textile fabrics generally; to our vases, our mosaics, our painted windows, our frescoed walls, our Academies, our Louvres, and our Vaticans. Even as we put sugar and spices into insipid dishes to gratify the gustatory nerves, whose sensibility was originally developed by the savor of tropical fruits, so do we put red, blue, and purple, into our pottery, our decoration, and our painting, to gratify the visual nerves, whose sensibility was originally developed by the rich tint of grapes and strawberries, star-apples and oranges.

And here again, as in the case of flowers, the feeling once aroused has found for itself new objects in the voluntary selection of beautiful mates—that is to say, of mates whose coloring gratified the rising delight in pure tints. The taste formed upon blossoms produced, by its reaction, crimson butterflies and burnished beetles, the sun-birds of the East, and the humming-birds of the West. So, too, the taste formed upon fruits produced, by a like reaction, parrots, cockatoos, toucans, birds-of-paradise, nutmeg-pigeons, and a thousand other tropical creatures of exquisite plumage and delicate form. As we mount up through the mammalian series, we scarcely come upon any hues brighter than dull-brown or tawny-yellow among the marsupials, the carnivores, the ruminants, or the thick-skinned beasts; but when we arrive at the seed-eating classes, such as the rodents, the bats, and the quadrumana, we find a profusion of color in many squirrels, flying-foxes, and monkeys; while Mr. Darwin does not hesitate in attributing to the same selective action the rosy cheeks, pearly teeth, blue eyes, and golden hair, of the human species.

Nor is it only in the choice of mates that the nascent taste for color displays itself. Even below the limits of humanity bright-hued objects afford a passing pleasure to more than one aesthetically-endowed species. Monkeys love to pull crimson flowers in pieces, dart in pursuit of brilliant tropical birds, and are attracted by the sight of red or yellow rags. Those queer little creatures, the bower-birds, carry the same feeling a step further by collecting fragments of brilliantly-colored objects to decorate their gaudy meeting-places. But, when we reach the race of man, we find the love of color producing far more conspicuous secondary results. The savage daubs his body with red or blue paint, and plants his garden with the scarlet hibiscus or the purple bougainvillia. Soon, with the rise of pottery and cloth-making, he learns the use of pigments and the art of dyeing. Next, painting proper follows, with all the decorative appliances of Egypt, India, China, and Japan, until at last our whole life comes to be passed in the midst of clothing and furniture, wall-papers and carpets, books and ornaments, vases and tiles, statuettes and pictures, every one of which has been specially prepared with dyes or pigments, to gratify the feeling originally derived from the contemplation of woodland berries by prehistoric man, or his frugivorous ancestors. And all these varied objects of civilized life may be traced back directly to the reaction of colored fruits upon the structure of the mammalian eye.

What a splendid and a noble prospect for humanity in its future evolutions may we not find in this thought that, from the coarse animal pleasure of beholding food, mankind has already developed, through delicate gradations, our modern disinterested love for the glories of sunset and the melting shades of ocean, for the gorgeous pageantry of summer flowers, and the dying beauty of autumn leaves, for the exquisite harmony which reposes on the canvas of Titian, and the golden haze which glimmers over the dreamy visions of Turner! If man, base as he yet is, can nevertheless rise to-day in his highest moments so far above his sensuous self, what may he not hope to achieve hereafter, under the hallowing influence of those chaster and purer aspirations which are welling up within him even now toward the perfect day!

1 - I trust that in the sequel the critical botanist will excuse me for having neglected the strict terminology of carpological science, and made no distinction between seeds and fruits. Some little simplification is absolutely necessary for general readers in this the most involved department of structural botany.

Early last year a paragraph went the round of the papers to the effect that a large female anaconda-snake, in the reptile-house at the Zoölogical Gardens, after a fast of a twelvemonth, had at length been induced to kill and swallow a duck. This very touchy and vindictive lady, it appears, had taken such grave offense at her capture in her South American home, and at her subsequent compulsory voyage to Great Britain, that she sulked persistently for a whole year, and invariably refused the keeper's most tempting offers of live rabbits or plump young pigeons. Month after month she lay passive in her cage, with her heart beating, her lungs acting, and all her vital functions proceeding with the usual slow regularity of snake-life; but not a mouthful of food did she attempt to take, and not a single fresh energy did she recruit from without to keep up the working of her animal mechanism. As I read this curious case of a genuine "fasting girl" in my "Times" one morning, the thought struck me forcibly—"Why, after all, should we expect her to feed? Why should she not go on forever without tasting a morsel? In short, why should we eat our dinner?" And I set myself to work at once to find out what was the general opinion of the unscientific public upon this important though novel question.

Singularly enough, I found that most people were content to eat their dinner in a very unreasoning and empirical way. They had ways been accustomed to dine daily from their childhood upward, they felt hungry at the habitual dinner-hour, and they sat down to their five courses with an unquestioning acceptance of the necessity for feeding to prevent starvation. But when I inquired why people who did not eat should starve, why they should not imitate the thrifty anaconda, and take one meal in a twelvemonth instead of three in a day, they appeared to regard my question as rather silly, and as certainly superfluous. Yet I must confess the query seems to me both pertinent and sensible; and it may be worth while to attempt some answer here in such language as can be understood of the people, without diving into those profound mysteries of formulæ and equations with which physicists love to becloud the subjects of their investigation.

A still more startling case than that of the anaconda will help to throw a little light upon the difficult problem which we have to solve. An Egyptian desert-snail was received at the British Museum on March 25, 1846. The animal was not known to be alive, as it had withdrawn into its shell, and the specimen was accordingly gummed, mouth downward, on to a tablet, duly labeled and dated, and left to its fate. Instead of starving, this contented gasteropod simply went to sleep in a quiet way, and never woke up again for four years. The tablet was then placed in tepid water, and the shell loosened, when the dormant snail suddenly resuscitated himself, began walking about the basin, and finally sat for his portrait, which may be seen of life-size in Mr. Woodward's "Manual of the Mollusca." Now, during those four years the snail had never eaten a mouthful of any food, yet he was quite as well and flourishing at the end of the period as he had been at its beginning.

Hence we are led to the inquiry—What is the actual function which food subserves in the human body? Why is it true that we must eat or we must die, while the snake and the snail can fast for months or years together with impunity? How do we differ from these lower animals in such a remarkable degree, when all the operations of our bodies so closely resemble theirs in general principle?

Everybody has heard it said that food is to men and animals what fuel is to a steam-engine. Everybody accepts this statement in a vague sort of way, but until the last few years nobody has been able really to explain what was the common feature of the two cases. For example, most people if asked would answer that the use of food is to warm the body, but this is really quite beside the question: because, in the first place, the use of fuel is not to warm the steam-engine, but to keep up its motion; and, in the second place, many animals are scarcely perceptibly warmer than the medium in which they live. Again, most people show in every-day conversation that they consider the main object of food to be the replacement of the materials of the body; whereas we shall see hereafter that its real object is the replacement of the energies which have been dissipated in working. Indeed, there is no more reason why the materials of an animal body should waste away of themselves, apart from work done, than there is for a similar wasting away in the case of a mineral body such as a stone. When an animal does practically no work, as in the instance of our desert-snail, his body actually does not waste, but remains throughout just as big as ever. So we must look a good deal more closely into the problem if we want to understand it, and not rest content with vague generalities about food and fuel. Such half knowledge is really worse than no knowledge at all, because it deludes us into a specious self-deception, and makes us imagine that we comprehend what in fact we have not taken the least trouble to examine for ourselves.

Let us begin, then, by clearly realizing what is the use of fuel to the steam-engine. Obviously, you say, to set up motion. But where does the motion come from? "From the coal," answers the practical man, unhesitatingly. "Well, not exactly," says the physicist, "but from the coal and the air together." All energy or moving power, as we now know, is derived from the union of two bodies which have affinities or attractions for one another. Thus, if I wind up a clock, moved by a weight, I separate the mass of lead in the weight from the earth, for which it has the kind of affinity or attraction known as gravitation. This attraction then draws together the weight and the earth; and, in doing so, the energy I put into it is given out as motion of the clock. Similarly with coal and air: the hydrogen and carbon of the coal have affinities or attractions toward the oxygen of the air, and when I bring them together at a high temperature (of which more hereafter) they rush into one another's embrace to form carbonic acid and water, while their energy is given off as heat or motion of the surrounding bodies. We might have whole minefuls of coal at our disposal; but if we had no oxygen to unite with it, the coal would be of no more use than so much earth or stone. In ordinary life, however, the supply of oxygen is universal and abundant, while the supply of coal is limited; and so, as we have to lay in coals, while we find the oxygen laid in for us, we always quite disregard the latter factor in our fires, and speak as though the fuel were the only important element concerned. Yet one can easily imagine a state of things in which oxygen might be deficient; and in a world so constituted it would have to be regularly laid on in pipes, like gas or water, if the people wished to have any fires.

All energy, then, is derived from the separation of two or more bodies having affinities for one another. So long as the bodies remain separate, the energy is said, in the technical slang of physics, to be potential; as soon as the bodies unite, and the energy is manifested as motion, it is said to be kinetic. But these words are rather mystifying to ordinary readers, and frighten us by their bigness and their abstract sound; so I shall take the liberty of altering them for our present purpose to dormant and active respectively, which are terms quite as well adapted to express the meaning intended, and not half so likely to land us in an intellectual cul-de-sac, or to envelop us in a logical fog. When we take a piece of coal and a lot of free oxygen, we possess energy in the dormant state. But though the oxygen has strong attractions for the carbon and hydrogen, they cannot unite, because their atoms do not come into close contact with one another, and because the two last-named substances are bound up in the solid form of the coal. We might compare their condition to that of a weight suspended by a string, which has strong attractions toward the earth, but cannot unite with it till we cut the string. Just analogous is our action when we apply a match to the coal. The heat first disintegrates or disunites little atoms of the hydrocarbons which make it up, and sets them in a state of rapid vibration among themselves. This vibration brings them into contact with the atoms of oxygen, which at once unite with them, causing a fresh development of heat, and a liberation of all the dormant energy, which immediately assumes the active form. The carbonic acid and water (or steam) thus produced fly up the chimney, carrying with them the little bits of unburned coal which we call smoke; and a current of fresh oxygen rushes in to unite with the fresh atoms of hydrogen and carbon which have been disengaged by the energy liberated from their fellows. So the process continues, till all the coal has been converted into carbonic acid and water—of course by the aid of a corresponding quantity of oxygen—and all the energy has been turned loose as heat upon the room in which we sit and upon the air outside.

In the case of an ordinary fire, where warmth is the single object we have in view, we only think of the heat, and disregard the other aspects of the process. But it is clear that an enormous amount of motion has also been set up by the energy of the free coal and oxygen, as exemplified by the draught up the chimney, and the numerous currents of air produced by its action within and without the room. Now, in a steam engine we deliberately make use of this motion for our own purposes by a specially devised mechanism. We allow the fire to heat and expand the water in the boiler, thus transferring to its molecules the separation which formerly existed between the atoms of the coal and the oxygen. Then we make the expanded water or steam push up the piston, and we connect the piston in turn with a crank which sets in motion the wheels, and so passes on the active energy to the mill, train, or ship which we desire to move, as the case may be. Thus the dormant energy of the coals and oxygen is liberated in the active state by their union, and is finally employed to effect movement in external bodies by the intermediation of the boiler. Even then the energy does not disappear: for energy, like matter, is indestructible; but it merely passes by friction as heat to that wonderful surrounding medium which we call ether, and is dissipated into the vast void of space, no longer recoverable by us, though quite as really existent as ever.

In what way, however, has all this to do with the reason for eating our dinners? Simply this: Men and other animals may be regarded from the purely physical point of view as a kind of conscious locomotive steam-engine, with whom food stands in the place of fuel, while the possible kinds of movement are infinitely more varied and specialized. I do not mean to advance any of those "automatic" theories which have been so current of late years. Whether they are true or false, they have nothing to do with our present subject. I only want to put in a plain light an accepted scientific truth. Men differ enormously from steam-engines in their possession of consciousness, wills, desires, pleasures, pains, and moral feelings; but they agree with them in the purely physical mechanism of their motor organs. A man, like a steam-engine, cannot move without his appropriate fuel; and if the fuel is not supplied, the fire goes out and the man dies. The exact manner in which the materials are utilized for keeping up this vital flame is the question to which we must now address ourselves.

Food-stuffs and coal agree essentially in the chief characteristics of their chemical constitution. Both consist mainly of hydrogen and carbon, and both possess energy in virtue of the fact that their affinities for oxygen are not satisfied. Water contains hydrogen, and carbonic acid contains carbon; but we can get no motion out of these, because in them the oxygen has already united with the atoms for which it had affinity, and the separation necessary for dormant energy has ceased to exist. But in bread, meat, potatoes, or coal, the hydrogen and carbon remain in their free state, ready to unite with oxygen whenever the chance is presented to them. All alike obtained their energy in the same way. The rays of sunlight falling upon the leaves of their original trees or plants separated the oxygen from the water and carbonic acid in the air, and built up the free hydrocarbons in their tissues. The energy which they thus drank in has remained dormant within them ever since: in the case of the bread for a few short months, in that of the coal for countless millions of geological cycles. But, however long it may have rested in that latent form, whenever an opportunity occurs the atoms will reunite with oxygen, and the energy will once more assume the active shape. There is really only one serious difference between coal and food, and that is that most foods contain another element, nitrogen, as well as carbon and hydrogen; and this nitrogen is an absolute necessity for the animal if it is to continue living. But there are good reasons for suspecting that nitrogen is not itself a fuel, being rather analogous in its nature to a match, and having for its business to set up the first beginnings of a fire, not to keep the fire going when it has once been lighted. So that this apparent difference of kind is really seen to be unimportant when we get to the bottom of the question.

The various matters which an animal eats consist of pure food-stuffs and of useless concomitant bodies: just as coal consists of pure fuel and of the useless mineral matter known as ash. When an animal eats his dinner, the process of digestion and assimilation takes place, and has the ultimate result of separating the pure food-stuffs from the useless concomitants. The latter bodies are rejected at once; but the food-stuffs are taken up by his veins, incorporated with the blood (which consists of food in different degrees of combustion), and used for building up the various portions of his body. Supposing the animal were a mere growing object like a crystal, with no work to perform and no consequent waste of material, the process would stop here, and the creature would wax bigger and bigger from day to day, without any alteration in place or redistribution of assimilated matter. But the animal is essentially a locomotive machine, and the purpose for which he has taken in his food is simply that he may use it up in producing motion. For a while he stores it away in his muscles, or lays it by for future use as fat; but its ultimate destination in every instance is just as truly to be consumed for fuel as is the case with the coal in the steam-engine.

The food, however, only gives us one half of the necessary materials for the liberation of dormant energy. Oxygen is needed to give us the other half. This oxygen we take in whenever we breathe. Animals like fishes or sea-snails obtain the necessary supply from the water by means of gills; for large quantities of oxygen are held in solution by water, and the needs of such comparatively sluggish creatures are not very great. With them a little energy goes a long way. Air-breathing animals like ourselves, on the other hand, need relatively large quantities of the energy-yielding gas in order to keep up the constant movements and high temperature of their bodies. Such creatures, accordingly, take in the oxygen by great inhalations, and absorb it in their lungs, where it passes through the thin membrane of the capillaries, or very tiny blood-vessels, and so mixes freely with the blood itself. Thus we have food, supplied to the blood by the stomach, the exact analogue of the coal in the engine; and oxygen, supplied to the blood by the lungs, the exact analogue of the draught in the engine. Whenever these two substances—the hydrocarbonaceous foods and the free oxygen—reunite, they will necessarily give out heat and set up active movements.

The exact place and mode of their recombination we cannot yet be said to fully understand. But even if we did, the details would be sufficiently dry and uninteresting to general readers; and we know quite enough to put the subject in a simple and comprehensible form before those who are willing to accept the broad facts without small criticism.

We may say, then, that the energies of the body are used up in two principal ways—automatically and voluntarily. The automatic activities are produced by the steady and constant oxidation of some portion of the food-stuffs in the blood and tissues. As this oxidation takes place, it sets up certain regular movements, which compose what is (very incorrectly) known as the vegetative life in animals. There are an immense number of these movements always going on within our bodies, quite apart from our knowledge or will. Such are the beating of the heart, with the consequent propulsions of blood through the system; the expirations and inspirations of the lungs, which supply us with the oxygen for carrying on these processes; the act of digestion and assimilation; and many other minor functions of like sort. But just as in the case of the steam-engine, so in the human or animal body, the union of the oxygen with the hydrocarbons, besides producing motion, liberates heat. This heat keeps the bodies of birds, quadrupeds, and human beings, which are all very active in their automatic movements, at a much higher temperature than the surrounding medium; while reptiles, fishes, and other "cold-blooded" creatures, having much less energetic motions of the heart and lungs—which of course betokens much less oxidation of food-stuffs—have bodies comparatively little different in warmth from the air or water about them. We thus see in part why it was that the anaconda and the desert-snail could go so long without food; though we cannot quite understand that question till we have examined the voluntary movements as well. It should be added that, though the latter class of actions also produce heat—as we all know when we walk about on a cold day to warm ourselves—yet the temperature induced by the automatic activities of the body alone is generally sufficient under normal circumstances to keep us comfortably warm. Thus, while we are asleep, only the actions of breathing and the beating of the heart continue; but the union of oxygen with the food-stuffs to produce these movements suffices as a rule to make bed quite hot enough for all healthy persons; and if we ever wake up cold after a good night's rest, we may be sure that our automatic activities are not what they ought to be.

The voluntary activities of the body are brought about in a slightly different manner. Directly or indirectly, they depend upon the union of oxygen and food-stuffs within the tissues of our locomotive muscles, the energy so liberated being made use of to bend or extend our bones or limbs in the particular way we desire. The muscles always contain (in a healthy and well-fed person) large quantities of such stored-up food-stuffs; and the blood supplies them from moment to moment with oxygen which may unite with the food-stuffs whenever occasion demands. But the union does not here take place regularly and constantly, as in the case of the automatic organs; it requires to be set up by an impetus specially communicated from the brain. That seat of the will is connected with the various voluntary muscles by the living telegraphic wires which we call nerves; and when the will determines that a certain muscle shall be moved, the nerves communicate the disturbance to the proper quarter, the necessary oxidation takes place, and the muscle contracts as desired. We do not quite know how the nerves and muscles perform these functions; but it is pretty certain that the nitrogen of our foods plays an active part in the process, and that, as I have already hinted, it acts in a manner somewhat analogous to that of a match. We may suppose, to put the matter in a familiar form, that the will sends down a sort of electric spark[1] to the muscle; and that this spark, lighting up the explosive nitrogen, causes an immediate union of the oxygen with the constituents of muscle, and so produces the visible movement.

Of course, voluntary actions, like automatic ones, liberate heat; but this heat is generally somewhat in excess of what is required for comfort, especially in hot weather. Lower animals, however, which have no fires and no artificial clothing, require it more than we do to keep us warm; and even we ourselves in wintry weather always feel chilly in the morning until we have had a good brisk walk to set up oxidation, and consequently liberate enough heat to make us comfortable.

Thus all motion, in the animal as in the steam-engine, depends upon the union of oxygen with food or body-fuel. It is true that in the animal body oxygen can unite directly with carbon and hydrogen without the necessity of a high temperature, which we saw was indispensable in the case of the coal, in order to bring the two sets of atoms within the sphere of their mutual attractions. But the difference is probably due to the different condition of the hydrocarbonaceous substances within the animal body; or else, as others conjecture, to the assumption by the oxygen of that peculiar state in which it is known as ozone. At any rate, the two processes do not disagree in any essential particular, being both cases in which free substances, possessing dormant energy by virtue of their separation and their affinity for one another, unite together, and in so doing liberate their energy as heat and visible motion.

There is, however, one important distinction of detail between the mechanism of a steam-engine and the mechanism of an animal body, which gives rise to many of the mistaken notions as to the use of food which we noticed above. In the engine, we put all the coal into the furnace, and burn it there at once; while the piston, cylinder, cranks, and wheels are not composed of combustible material, but of solid iron. In the animal body, on the other hand, every muscle is at once furnace, boiler, and piston; it consists of combustible materials, which unite with oxygen in the tissues themselves, and set up motion within the muscle of which they form a portion. The case is just the same as though the joints of an engine, instead of being quite rigid, were composed of hollow India-rubber and whalebone, with iron attachments; were then filled with coal, oxygen, and water, and possessed the power of burning up these materials internally and setting up motions in the India-rubber tubings. Hence the materials in the muscles are always undergoing change. The carbon and hydrogen which have united with the oxygen are perpetually forming carbonic acid and water; and, as [2] these have lost or given up all their energy, they are naturally of no more use to the body than the similar carbonic acid and steam which fly up the draught are of use to the engine. Accordingly, they are taken up by the stream of blood as it passes, separated from the useful components of that compound liquid by an appropriate organ, and rejected from the body as of no further service.

But their place in the muscle must once more be supplied by fresh energetic materials; and these materials are brought to it by the selfsame blood which removes the de-ënergized waste products. And now we begin to see why we must eat our dinners or starve. Every time our heart beats, every time our lungs draw in a breath, a certain amount of matter in the tissues of the muscles which produced those motions undergoes oxidation, and is carried off in the oxidized form to be cast out of the body as waste. Every new pulsation or breath requires a certain new quantity of energetic material, both as food-stuffs and as oxygen; and hence we must supply the one from the stomach and the other from the lungs if we wish to keep the mechanism going. The store of hydro-carbonaceous matters laid by in the body is generally considerable in well-fed persons; for, besides the contents of the muscles themselves, we have usually a large reserve fund in the shape of fat, ready to be utilized when occasion arises. Hence, we can get along for a very short time, if necessary, without food; because we can fall back, first upon the fat-reserve, and then upon the muscles and tissues, for energetic materials. But after a time the ceaseless beating of the heart and movement of the lungs will use up all the available matters, and the blood will cast off the oxidized product and excrete it from the body; till at last no more materials are forthcoming, the whole contents of the tissues have been oxidized and got rid of, and the heart and lungs must perforce cease to act, in which case the unhappy victim is said to have died of starvation. As regards the supply of oxygen, on the other hand, we are very much more restricted in our power of endurance; for we have no large store of this necessary for combustion laid by in our bodies, and if the supply be cut off for a single moment (as by compressing the throat or suffocating with carbonic acid) the heart and lungs must cease at once to act, and death takes place immediately. For of course death, viewed on its purely physical side, means the cessation of that set of activities which results from the union of oxygen with the food-stuffs in the body.

By this time I hope the reader can see quite clearly what is the necessity for eating his dinner. If we are to live, we must keep up the cycle of our bodily activities, and especially those two fundamental ones, the breathing of the lungs and the beating of the heart. In order to do this, we must supply the muscles employed with the two energy yielding substances, oxygen and hydrocarbons. The supply of oxygen must be continuous; in other words, we must never for a moment leave off breathing; but the supply of hydrocarbons may be intermittent, though it must be sufficient on the whole to balance waste. We must not regard the object of food, however, as being merely to build up the matter of the body; we must rather consider it as intended to recruit the energies of the body. The more active any creature is, both in its automatic and its voluntary movements, the greater will be the amount of hydrocarbons consumed or used up in its muscles, and the greater, consequently, the amount of food and oxygen which it will require to make up the loss. The tiny humming-bird will need far more food in a year than the great anaconda with which we began our discourse: because the humming-bird has a rapidly moving heart and lungs, while the cold-blooded snake respires and circulates slowly; and the humming-bird darts about perpetually at lightning-speed from flower to flower, while the snake lies coiled up motionless in its blanket from year's end to year's end, or only comes out sleepily now and then to swallow the food which will keep up its vital actions through another long and lazy fast.

The desert-snail, however, can endure much longer without food than even the anaconda, because, like so many other mollusca, it can hibernate. This process of hibernation consists in the inducement of a state during which the heart ceases to beat, respiration is suspended, and the animal can hardly be said to live at all. But when warmth and moisture are once more applied, the heart recommences its action, the lungs or gills quicken their movements, voluntary locomotion ensues, and the creature sets out again on the quest for food. Something analogous occurs in the case of the bear, the dormouse, and other hibernating quadrupeds; but in these instances the vital functions continue much more in their ordinary state, and are kept up by the supply of fat which is dissolved by the blood, and consumed in effecting the necessary automatic actions. The bear, which goes to sleep in the autumn as sleek and plump as a prize pig, wakes up in the spring a poor, lean wretch, with only just flesh enough to cover his bones, and carry him off in search of fresh food. The much more complicated mechanism of the higher animals requires to be kept always in action; it can not cease almost entirely, like that of the snail, and then revive again when circumstances become more favorable. Hence hibernating mammals must lay by fat during the summer to keep their principal organs at work during the long winter fast. Yet, even among human beings, cases of "trance" or "suspended animation" occasionally occur, during which the cycle of vital actions almost entirely ceases to all appearance for a considerable time, and then begins again on the application of some external or internal stimulus—which latter may be not unaptly compared to the slight shaking which we sometimes give a watch or clock to set it going when stopped by a momentary impediment. Persons recovered from drowning, in whom the cessation of action has been quite sudden and has not affected the structure of their organs, are often thus restored by the judicious use of rubbing and alcohol.

The camel presents a more interesting phenomenon in his well-known humps. These protuberances consist really of reserve-stores of fat, which the camel uses, not only for keeping up the action of his heart and lungs, but also for producing locomotion in his frequent enforced fasts among the deserts of Arabia or India. The humps dwindle away as he marches, in a manner exactly similar to that of the bear's fat during his hibernation, only of course much more rapidly, as they have so much more work to perform.

Finally, it may appear strange that the small amount of food we eat should suffice to carry our large and bulky bodies through all the varied movements of the day. But this difficulty disappears at once when we recollect how large an amount of energy can be laid by dormant in a very small piece of matter. A lump of coal no bigger than one's fist, if judiciously employed, will suffice to keep a small toy-engine at work for a considerable time. Now, our food is matter containing large amounts of dormant energy, and our bodies are engines constructed so as to utilize all the energy to the best advantage. A single gramme of beef-fat, if completely burned (that is, if every atom unites with oxygen), is capable of developing more than 9,000 heat-units; and each such heat-unit, if employed to perform mechanical work, is capable of lifting a weight of one gramme to a height of 424 metres; or, what comes to the same thing, 424 grammes to a height of one metre. Accordingly, the energy contained in one gramme of beef-fat (and the oxygen with which it unites) would be sufficient to raise the little bit of fat itself to a height of 3,816 kilometres, or about as high as from London to New York. Again, it may seem curious that the food eaten by the anaconda in South America, and stored up in its tissues, should suffice to keep up the action of its heart and lungs for so many months. But then we must remember that it performed very few other movements, most probably, during all that time; and if we think how small an amount of energy we expend in winding up an eight-day clock, and how infinitesimal a part of our dinner must have been used up in imparting to it the motion which will keep it swinging and ticking for one hundred and ninety-two hours, we can easily understand how the large amount of stored-up energy in the snake's muscles might very well serve to keep up its automatic actions for so long a time.

There are five hundred other little points which this mode of regarding our bodies at once clears up. It shows us why we are warmer after eating a meal, why cold is harder to endure when we are hungry, why we need so little food when we are lying in bed inactive, and so much when we are taking a walking tour or training for a boat-race, why cold-blooded animals eat so rarely and warm-blooded creatures so often, why we get thin when we take too little food, and why we lay on fat when we take too little exercise. But these and many other questions must be passed over in silence, or left to the reader's discrimination, lest I should make this paper tediously long. It must suffice for the present if I have given any of my readers a more rational reason in future for eating their dinners. To be sure, Nature herself has admirably provided that even the most unscientific person should find sufficient internal conviction as to the desirability of dining without the aid of extraneous exhortation; but it is at least some comfort to know that so universal and so unreasoning a practice is not altogether an unreasonable one as well.

1 - I am speaking quite metaphorically and popularly, and do not mean to imply adhesion to the electrical rather than to the isomeric theory of nervous conduction.

2 - I purposely simplify and omit details, so as to give the reader a graphic and comprehensive picture of the central facts. So long as essentials are not distorted, a good diagram is far better for educational purposes than an accurate facsimile.

A PROBLEM IN HUMAN EVOLUTION

"Hardly any view advanced in this work," says the illustrious author of the "Descent of Man," "has met with so much disfavor as the explanation of the loss of hair in mankind through sexual selection." Indeed, the friends and foes of Mr. Darwin's great theories have been equally ready, the one party to disclaim and the other party to ridicule the account which the founder of modern philosophic biology has given of the process whereby man, as he supposes, gradually lost the common hairy covering of other mammalia. Mr. Wallace, with all his ability and ingenuity, finds it necessary to call in the aid of a deus ex machina to explain the absence of so useful and desirable an adjunct; for he believes that natural selection could never have produced this result, and he therefore feels compelled to put it off upon "some intelligent power," since he denies altogether the existence of sexual selection as a vera causa. Mr. J. J. Murphy, in his recently published revision of "Habit and Intelligence," has taken up the same ground with a more directly hostile intent; and Spengel has also forcibly given expression to his dissent on the plea of inadequate evidence for the supposed preference. It seems highly desirable, therefore, to prop up Mr. Darwin's theory by any external supports which observation or analogy may suggest, and if possible to show some original groundwork in the shape of a natural tendency to hairlessness, upon which sexual selection might afterward exert itself so as to increase and accelerate the depilatory process when once set up.

The curious facts for which we have to account are something more than the mere general hairlessness of the human species. In man alone, as Mr. Wallace clearly puts the case, "the hairy covering of the body has almost totally disappeared; and, what is very remarkable, it has disappeared more completely from the back than from any other part of the body. Bearded and beardless races alike have the back smooth, and even when a considerable quantity of hair appears on the limbs and breast, the back, and especially the spinal region, is absolutely free, thus completely reversing the characteristics of all other mammalia." When we consider the comparatively helpless condition to which man has been thus reduced, as well as the almost universal human practice of substituting artificial clothing, derived from the skins or wool of other animals, for the natural apparel which the species has so unaccountably lost, it does not seem surprising that even Mr. Wallace should be staggered by the difficulty, and should fall back upon an essentially supernatural explanation.

The great key to the whole problem lies, it would seem, in the fact thus forced upon our attention, that the back of man forms the specially hairless region of his body. Hence we must conclude that it is in all probability the first part which became entirely denuded of hair. Is there any analogy elsewhere which will enable us to explain the original loss of covering in this the normally hairiest portion of the typical mammalian body? The erect position of man appears immediately to suggest the required analogy in the most hairless region of other mammals.

Almost all animals except man habitually lie upon the under surface of the body. Hence arises a conspicuous difference between the back and the lower side. This difference is seen even in lizards, crocodiles, and other reptiles, among which, as a rule, the tegumentary modifications of the under surface are much less extended and less highly differentiated than those of the upper. It is seen among birds, which usually have the plumage far less copious on the breast than on the back. But it is most especially noticeable in mammals, which have frequently the under side almost entirely bare of hair, while the back is covered with a copious crop. Now, it would seem as though this scantiness of natural clothing on the under side were due to long continued pressure against the ground, causing the hair to be worn away, and being hereditarily transmitted in its effects to descendants. We are, therefore, led to inquire whether all parts of the mammalian body which come into frequent contact with other objects are specially liable to lose their hair.

The answer seems to be an easy one. The soles of the feet in all mammals are quite hairless where they touch the ground. The palms of the hands in the quadrumana present the same phenomenon. The knees of those species which frequently kneel, such as camels and other ruminants, are apt to become bare and hard-skinned. The callosities of the Old-World monkeys, which sit upon their haunches, are other cases in point; but they do not occur among the more strictly arboreal quadrumana of the American Continent, nor among the lemurs, for the habits of these two classes in this respect are more similar to those of ordinary mammals. On the other hand, the New World monkeys possess a prehensile tail, with which they frequently swing from bough to bough or lower themselves to the ground, and in these creatures, says Cuvier, "la partie prenante de la queue est nue en dessous." Wherever we find a similar organ, no matter how widely different may be the structure and genealogy of the animals which possess it, we always find the prehensile portion free from hair. This is the case with the marsupial tarsipes, with many rodents, and above all with the opossum, which uses its tail quite as much as any monkey uses its hands. Accordingly, its surface is quite bare from end to end, and in some species scaly — a fact which is rendered more comprehensible when we remember that the young opossums are carried on their mother's back, and hold themselves in that position by curling their tails around hers.

A few more special facts help to bear out the same generalization. In the gorilla, according to Du Chaillu, "the skin on the back of the fingers, near the middle phalanx, is callous and very thick, which shows that the most usual mode of progression of the animal is on all fours and resting on the knuckles." The ornithorhyncus has a flat tail, on which it leans for support, and this, says Mr. Waterhouse, "is short, depressed, and very broad, and covered with coarse hairs; these, however, are generally worn off on the under side of the tail in adult or aged individuals, probably by the friction of the ground." The toes of the very large fore-feet, used in burrowing, are also naked, as are the similar organs in the mole and many other creatures of like habit. The beaver likewise uses his tail as a support, flaps it much in the water, and is said, perhaps not quite erroneously, to employ it as a trowel in constructing his dams; and this tail is entirely devoid of hair, being covered instead with a coat of scales. We can hardly avoid being struck in this instance, as in that of some seals' and sea-lions' flappers, with the analogy of the penguin's wings, which are employed like fins in diving, and have undergone a similar transformation of their feathers into a scale-like form. In the ground kangaroos, which use the tail as a support trailing behind them on the ground, that organ is again only slightly covered with coarse hairs, almost entirely wanting on the extremity of the under surface; but in the tree-kangaroos, which carry the tail partly erect, it assumes a bushly and ornamental appearance. Like differences occur between the rats and mice on the one hand and the squirrels on the other. In those monkeys which, like Macacus brunneus, sit upon their tails, that organ is also bare. To multiply further instances would only prove tedious.

Again, when we look at the only mammals besides man which have denuded themselves of their hairy covering, we find that a great majority of them are water-frequenters. The most completely aquatic mammals, like the whales, porpoises, dugongs, and manatees, though differing widely in structure, are alike in the almost total absence of hair, while the hippopotamus is likewise a smooth-skinned animal. Now, the friction of water is of course far stronger than that of air, and it would seem to have resulted in the total depilation of these very aquatic species. Other less confirmed water-haunters, such as seals and otters, have very close fur, which scarcely at all retards them in their movements when swimming. The elephant and rhinoceros are, indeed, difficult cases to explain; but of course it is not necessary to suppose that no other cause save that which we are considering can ever produce hairlessness. It will be enough if we can show that the cause actually under examination does with reasonable certainty bring about such an effect.

If, then, the portion of animals which generally comes in contact with the ground or other external bodies acquires in this manner a hairless condition — shown alike in hands, feet, tail, and belly — what will be the result upon animals which are gradually acquiring the erect position? Of this we can obtain an almost complete series by looking first at the beaver, which rests upon its scaly tail alone; then at the baboons, which rest upon the naked callosities on their haunches; thirdly, at the gorilla; and, last of all, at mankind.

The gorilla, according to Professor Gervais, is the only mammal which agrees with man in having the hair thinner on the back, where it is partly rubbed off, than on the lower surface. This is a most important approach to a marked human peculiarity, and is well worthy of investigation. "I have myself come upon fresh traces of a gorilla's bed on several occasions," says Du Chaillu, "and could see that the male had seated himself with his back against a tree-trunk. In fact, on the back of the male gorilla there is generally a patch on which the hair is worn thin from this position, while the nest-building Troglodytes calvus, or bald-headed nshiego, which constantly sleeps under its leafy shelter on a tree-branch, has this bare place on its side, and in quite a different way. . . . When I surprised a pair of gorillas," he observes elsewhere, "the male was generally sitting down on a rock or against a tree." Once more, in a third passage he writes: "In both male and female the hair is found worn off the back; but this is only found in very old females. This is occasioned, I suppose, by their resting at night against trees, at whose base they sleep." And, when we inquire into the difference between the sexes thus disclosed, we learn that the female and young generally sleep in trees, while the male places, himself in the position above described against the trunk.

The gorilla has only very partially acquired the erect position, and probably sits but little in the attitudes common to man. But if a developing anthropoid ape were to grow more and more upright in his carriage, and to lie more and more upon his back and sides, we might naturally expect that the hair upon those portions of his body would grow thinner and thinner, and that the usual characteristics of the mammalia as to dorsal and sternal pilosity would be completely reversed. This is just what has probably happened in the case of man. In proportion as he grew more erect, he must have lain less and less upon his stomach, and more and more upon his back or sides. For fully developed man, with the peculiar set of his neck, face, and limbs, it is almost impossible to rest upon his stomach. On the other hand, all savage races lie far more upon their backs than even Europeans with their sofas, couches, and easy-chairs; for the natural position of savage man during his lazy hours is to stretch himself on the ground in the sun, with his eyes closed, and with his back propped, where possible, by a slight mound or the wall of his hut. Any person who has lived much among negroes or South Sea Islanders must have noticed how constant is this attitude with men, women, and children, at every stray idle moment.

Nor must we forget the peculiar manner in which human mothers must necessarily have carried their infants from a very early period in the development of our race. During the first eighteen months of life the human infant must always be held, or laid, more or less upon its back; and this position will probably tend to check the development of hair upon the dorsal and lateral regions.

Next, let us ask what is the actual distribution of hair upon the body of man. Omitting those portions where the ornamental use of hair has specially preserved it, the most hairy region is generally, so far as my observations go, the fore part of the leg or shin. Obviously this is a region very little likely to come in contact with external objects. On the other hand, the most absolutely hairless places are the palms of the hands and the soles of the feet, after which come the elbows, and at a long interval the knees and knuckles. The back is very hairless, and so are the haunches. But the legs are more hairy than the body, both in front and behind, though less hairy on the calf than on the shin. Now, it will be obvious that both by day and night we rest more upon our backs and haunches than upon our legs, the latter being free when we sit down on a chair or bench, doubled in front of us when we squat on the ground (the normal position of savages), and thrown about loosely when we lie down. Especially might we conclude that this would be the case with early races, unembarrassed by the weight of bedclothes. As for the arms, it is noticeable that they still retain the ordinary mammalian habit in being hairier on the back than on the front; and this also is quite in accordance with our present suggestion, because the same differentiating causes have not worked upon the arm as they work upon the back and legs. The peculiar position of the anterior extremities in man, together with the erect posture, makes the arms come much more frequently into frictional contact with the body or clothing on their inner than on their outer surface. Hair grows most abundantly where there is normally least friction, and vice versa. As for the hair which frequently appears upon the chest of robust Europeans and others, I shall return to that point at a later stage. It may be noted, however, that while the first joint of the fingers is hairy, the second joint, answering to the callosity of the gorilla, is generally bare.

As man, then, gradually assumed the erect attitude and the reversed habits of sitting and lying down which it necessarily involves, it seems to me that he must have begun to lose the hair upon his back. But such a partial loss will not fully account for his present very hairless condition over the whole body (with trifling exceptions) in the average of all sexes, races, and ages. For this further and complete denudation I think we must agree with Mr. Darwin in invoking the aid of sexual selection, especially when we take into consideration the ornamental and regular character of the hairy adjuncts which man still retains.

In the first place, we have external reasons for believing that sexual selection has produced similar results elsewhere, acting upon a like basis of natural denudation. For among the mandrills and some other monkeys the naked callosities, originally produced, as is here suggested, by physical friction, have been utilized for the display of beautiful pigments; and Mr. Bartlett informed Mr. Darwin that as the animals reach maturity the naked surfaces grow larger in comparison with the size of the body. When we look at the great definiteness and strange coloring of these bare patches we can hardly doubt that they have been subjected to some such selective process.

But if man once began to lose the hair over the whole of his back, shoulders, and haunches, as well as more partially upon his sides, legs, and arms, he would soon present an intermediate half-hairy appearance which is certainly very ludicrous and shabby-looking. Why this middle stage should displease us, it might be rash to guess; yet one may remember that as a rule throughout the mammalia a partially hairless body would be associated with manginess, disease, and deformity. At any rate, it seems to be the fact that, when animals once begin losing their hair, they go on to lose it altogether. One may well believe that among our evolving semi-human ancestors those individuals which had most completely divested themselves of hair would be the most attractive to their mates; and these would also on the average be those which had most fully adopted the erect attitude with its accompanying alterations of habit. Thus natural selection would go hand in hand with sexual selection (as I believe it always does), those anthropoids which most nearly approached the yet unrealized standard of humanity being most likely to select one another as mates, and their offspring being most likely to survive in the struggle for life with their less anthropoid competitors. [1] It does not seem probable, to me at least, that a naturally hairy species would entirely divest itself of its hair through sexual selection, especially as the first steps of such a process could hardly fail to render it a mongrel-looking and miserable creature; but it seems natural enough that, if the original impulse was given by a physical denudation, the influence of sexual selection would rapidly strengthen and complete the process. Indeed, if a hairy animal once began losing its hair, the only beauty which it could aim at would be that of a smooth and shiny naked black skin.

Woman is the sex most affected in mankind by sexual selection, as has been often abundantly shown. Hence we should naturally expect the denudation to proceed further in her case than in that of man. Especially among savage and naked races we should conclude that hairlessness on the body would be esteemed a beauty; and Ave find as a matter of fact that most such races have absolutely smooth and glistening skins. But in Europe men often develop hair about the chest and legs, though not upon the back and shoulders, while women seldom or never do so. Here we see that the hair reappears in the less differentiated male sex rather than in the more differentiated females, with whom sexual selection has produced greater effects; while it also reappears only on those parts where the original denudating causes do not exert any influence. Similarly, the smooth-bodied negroes, transported to America, and subjected at once to a change of conditions and to circumstances which would render sexual selection impossible as regards the hairlessness of the body, rapidly redevelop hair upon the chest. For we must remember that sexual selection can only act in this direction while a race remains wholly or mainly naked. Clothing, by concealing the greater part of the skin, necessarily confines the selective process to features, complexion, and figure.

As to the poll, beard, whiskers of certain races, we must believe that they are the result of selective preferences acting upon general tendencies derived from earlier ancestors, and perhaps aided in the first-mentioned instance by natural selection. The comparative definiteness of these hairy patches, as of the callosities in the monkeys, stamps them at once as of sexual origin. The poll is probably derived by us from some of our anthropoid ancestors, as crests of hair frequently appear upon the heads of the quadrumana. But as man gradually became more erect and less forestine, as he took to haunting open plains and living more in the sunlight, the existence of such a natural covering, as a protection from excessive heat and light upon the head, would doubtless prove of advantage to him; and it might, therefore, very possibly be preserved by natural selection. Certainly it is noticeable that this thick mat of hair occurs in the part of his body which the erect position most exposes to the sunlight, and is thus adaptively analogous to the ridge of hair which runs along the spine or top of the back in many quadrupeds, and which is not visible in any quadrumanous animal that I have examined. The beard also bears marks of a quadrumanous origin, as Mr. Darwin has shown; but its varying presence or absence in certain races affords us a good clew to the general course of evolution in this particular. For among the bearded races a fine and flowing beard is universally admired; while among the beardless races stray hairs are carefully eradicated, thus displaying the same aversion to the intermediate or half-hairy state which, as I suppose, has been mainly instrumental in completely denuding the body of man. Certainly it is a fact that while we can admire a European with a full and handsome development of hair upon the chin and lip, and while we can admire an African or a North American Indian with a smooth and glossy cheek, we turn with dislike from thin and scanty hair either in a European, a negro, or an Asiatic. It seems to me that in every case the general æsthetic feeling of the whole human race is the same; but that in one tribe circumstances have made it easier to produce one type of beauty, while in another tribe other conditions have determined the production of another type. Thus, in a negro, a very black and lustrous skin, clear bright eyes, white teeth, and a general conformity to the normal or average negro features are decidedly pleasant even to Europeans when once the ordinary standard has become familiar;[2] while in a European the same eyes and teeth are admired, but a white skin, a rosy complexion, and moderate conformity to the ideal Aryan type are demanded. Each is alike pretty after its own kind, though naturally the race to which we each ourselves belong possesses in most cases the greatest attractiveness to each of us individually.

Of course, both in the beard of man, and in the general hairiness of his body, as compared with woman, allowance must be made for that universal tendency of the male to produce extended tegumentary modifications, which, as Mr. Wallace has abundantly shown, depends upon the superior vigor of that sex. Yet the period when the beard first shows itself and the loss of color in the hair of both sexes after the reproductive period is past clearly stamp these modifications as sexual in origin.

It must be remembered also, in accounting for the general loss of hair on both back and front of the body, that the older ancestral heredity would tend to make the chest bare, and the newer acquired habits would tend to produce like results upon the back. "In the adult male of the gorilla," says Du Chaillu, "the chest is bare. In the young males which I kept in captivity it was thinly covered with hair. In the female the mammæ have but a slight development and the breast is bare." All this helps us to see how the first steps in the sexually selective process might have taken place, and also why the trunk is on the whole more denuded than the legs. As for the exceptional fact that the arms are hairier on the back than in front, besides the functional explanation already given, we must recollect that the anthropoid apes have long hair on the outer side of the arms, which has probably left this slight memento of its former existence on the human subject. Eschricht has pointed out the curious fact that alike in man and the higher quadrumana this hair has a convergent direction toward the point of the elbow, both from above and from below.

Finally, it may be noted that the hairless condition of man, though apparently a disadvantage to him, has probably been indirectly instrumental in helping him to attain his present exalted position in the organic scale. For if, as is here suggested, it originally arose from the reactions of the erect attitude, it must have been associated from the first with the most human-like among our ancestors. Again, if it was completed by sexual selection, it must also have been associated with the most æsthetic individuals among the evolving species. And if, as we have seen reason to believe, these two qualities would tend to accompany one another, then this slight relative disadvantage would be pretty constantly correlated with other and greater advantages, physical and intellectual, which enabled the young species to hold its own against other competing organisms. But, granting this, the disadvantage in question would naturally spur on the half-developed ancestors of man to seek such artificial aids in the way of clothing, shelter, and ornament, as would ultimately lead to many of our existing arts. We may class the hairlessness of man, therefore, with such other apparent disadvantages as the helpless infancy of his young, which, by necessitating greater care and affection, indirectly produces new faculties and stronger bonds of union, and ultimately brings about the existence of the family and the tribe or nation. And if we look back at the peculiarities which distinguish placental from implacental mammals, the mammalia generally from birds, and birds from reptiles, we shall see that in every case exactly similar apparent disadvantages have been mainly instrumental in producing the higher faculties of each successive vertebrate development. Hence it would seem that the hairless condition of man, instead of requiring for its explanation a special intervention of some supernatural agent, is strictly in accordance with a universal principle, which has brought about all the best and highest features of the most advanced animal types through the unaided agency of natural selection.

1 - On the advantages which man or his half-developed ancestor derived from the erect or semi-erect position, see Darwin, "Descent of Man," p. 53.

2 - The mutilations of the face and other parts, which often make savages so ugly in our eyes, though not in their own, are due, as Mr. Herbert Spencer has shown, not to æsthetic intentions, but to originally subordinative practices, as marks of subjection to a conquering king or race

"PLEASED WITH A FEATHER"

A murky London winter afternoon is not exactly a good opportunity for the pursuit of natural history. The snow lies thick on the pavement outside, half melted into muddy slush; while the fog penetrates through the cracks in the woodwork, and the sun struggles feebly athwart the thick yellow sheet which shuts off his rays from the lifeless earth. If I wish to go on a botanical or entomological excursion to-day, I must perforce content myself with a "Voyage autour de ma Chambre." So I rise listlessly from my easy-chair; perambulate the drawing-room in a sulky mood; peer at the Japanese fans on the mantel-shelf; rearrange for the twentieth time those queer little pipkins we brought on our last vacation ramble from Morlaix; pull about my wife's old Chelsea in a savage fit of tidiness; and finally relapse upon the sofa with a fixed determination to be inconsolably miserable for the rest of the day. Evidently I am suffering from that mysterious British epidemic, the spleen, and I may be shortly expected to plunge incontinently over Waterloo Bridge.

Meanwhile, I find a momentary solace in the Indian cushion which lies under my head. A feather is just pushing its sharper end through the morocco-leather groundwork, between those gorgeous masses of gold, silver, and crimson embroidery; which feather I forthwith begin to egg out, by dexterous side pressure, with admirable industry, worthy of a better cause. My wife, looking up from her crewels, mutters something inarticulate about someone who finds some mischief still for idle hands to do; but her obdurate husband pretends inattention, and finally succeeds in catching the feather-end between his finger and thumb. Now that I have successfully pulled it out, I begin to examine it closely, and bethink myself of how, in brighter summer weather, I dissected a daisy for the benefit of such among the readers of the "Cornhill Magazine" as honored me with their kind attention. I shall take a closer look at this feather, and see if it, too, may not serve as a text for a humble lay-sermon concerning the nature and development of feathers in general, and the birds or human beings who wear them.

For the interesting point about a feather is really this, that it grew. It was not made in a moment, like a bullet poured red-hot into a mold: its little airy plumes, branched like a fern into tiny waving filaments, were developed by slow steps, piece after piece, and spikelet after spikelet. And what is true of this particular bit of down which I hold in my fingers, trembling like gossamer at every breath and every pulse, is also true of plumage as a whole in the history of animal evolution. To my mind that great fact, that everything has grown, throws a fresh and wonderful interest into every little object which we can pick up about our fields or our houses. The old view of creation, which represented it as single and instantaneous, made each creature or each organ seem like a mere piece of molded mechanism, with no history, no puzzle, and no recognizable relation to its like elsewhere. But the new view, which represents creation as continuous, progressive, and regular, teaches lis to see in every species or every structure a result of previous causes, an adaptation to preexisting needs. Thus we are enabled to find in a flower, a fruit, or a feather, innumerable clews which lead us back to its ultimate origin, and give delightful exercise to our intelligence in tracing out the probable steps by which this complex whole has been produced.

I often figure to myself the difference between the two ways of regarding natural objects, by means of the initial letters in an ordinary volume, and the initial letters which Mr. Linley Sambourne draws for us so cleverly in "Punch." Look at the big O of a newspaper leader — it is just a mass of metal, poured into a circular or oval type. But look at the big O which the ingenious artist tricks out for us with social allusions or political innuendoes, and what a world of amusement you will find if you take the trouble to spell out all its quaint devices! See how every curl has some playful hit at a noble lord or an honorable member; how every detail smiles with gentle satire at some passing event or some universal topic. Not a touch but has a meaning for those who will seek it; not a careless little smudge in the corner but brims over with deep purpose and infinite wealth of covert mirth. So it is, I think, with flowers, fruits, or feathers, when once we have learned to look for their hidden hints. This little twist points back to some strange fact in the past history of the species; that unobtrusive spur or knob is the clew to whole volumes of botanical or zoological lore. Not a detail but tells of the origin and development of the whole; not a tuft, a spot, or a streak but teems with information for the seeker who has found out the method of seeking aright.

Again, to vary our simile, let us visit some ancient British earthwork or Roman camp. If we go as mere rustics, we see in it all nothing more than a broken ridge of earth on the summit of a rolling down. We are not even sure whether it is really the handiwork of man, or some queer natural formation like the Devil's Dike, the Giant's Causeway, and the parallel roads of Glen Roy. But if we go under the guidance of some skilled archaeologist, what a flood of light he is able to throw over its history and its meaning! This row of strongholds, he tells us, formed the frontier line, say between the Welsh of Dorset and the Welsh of Devon. Here the Durotriges and Damnonii, the men of the water-vale and the men of the hills, faced one another from their opposite heights. Sweep round your eye in a semicircle along this series of points, overhanging the valley of the Axe, and you will find every higher summit crowned with a "castle," a rude earthwork raised by the men whom our fathers drove out of the land. That was their Balkan or Suleiman line, their cordon of border forts, their row of beacons to announce the approach of the hostile hill-men on the war-trail against their homes. Then our antiquary would turn to the work itself, and would point out the various parts, the mode of defense, the simple tactics of those primitive Vaubans. Or else he would show us the Roman detail of the later encampment; the square scar that marked the prætorian quarters; the regular succession of gates and defenses. All this he would tell us from the bare inspection of the existing remains, reconstructing the lost history from his stored-up knowledge of like instances elsewhere.

But I am wandering sadly from my London room and my little feather, this wintry afternoon. Let me look at it once more, and try to realize, in like manner, the story involved in its downy vans.

In the first place, this feather, as an anatomist would tell us, is "a dermal modification" — in other words, an altered bit of the skin. Every part of a plant or animal undergoes changes, our modern teachers say, just in accordance with the external influences which affect it. But the skin of an animal is naturally exposed to many more such surrounding agencies than its internal organs. Accordingly, we find that no structure exhibits such strange variations as the skin. Besides the regular modifications which we see in the scales or horny plates of fishes, the smooth coats or solid shells of reptiles, the feathers of birds, and the hair of mammals, numerous other minor peculiarities occur in almost every species. Such are the horns of cows and goats, the spike of the rhinoceros, the beaks, nails, claws, hoofs, and talons of beasts or birds, and the tail-plumes, ruffs, lappets, crests, and ornamental adjuncts of all the more aesthetic animals. In no class are these variations in the external covering more conspicuous than among the biped tribe whose spoils I am now holding in my hand as the text for our afternoon's discourse.

How birds first came to be winged and feathered we can hardly say as yet. To be sure, most of us have seen a picture, at least, of that strange oolitic monster, the pterodactyl, a saurian with a head like a crow, but having the fore-part protracted into long jaws, fitted with teeth not very dissimilar from those of a crocodile; while its legs were supplied, apparently, with a membrane, by whose aid the creature probably flew about in the same manner as a bat. These real flying dragons recall in many points the appearance of a bird, especially in the skull and the position of the eyes. Moreover, Professors Marsh and Huxley have shown that the earliest fossil birds resemble the pterodactyl and other reptiles in many important peculiarities of structure, far more than their modern representatives. Some of them even possess teeth set in their jaws after a reptilian fashion. Though the evidence still remains very fragmentary, we may regard it as probable that birds are descended from some early reptilian form, more or less like the peterodactyl, if not actually from that partially-winged saurian itself. But perhaps it is premature to build with any confidence upon such dubious ground; and we may consequently accept the earliest birds on their own responsibility, without inquiring too curiously into their antecedents, or compelling them to produce a genealogical table of their ancestry.

The essential characteristic of a bird consists in the fact that it is a flying animal; and feathers are the kind of skin-covering best adapted to its special manner of life. In their nature and mode of development, feathers closely agree with the hair of mammals; but the differences between them are all of a sort which fit the bird for its aërial existence. We see this fact very clearly if we look at the instance of those birds which do not fly. Running species, such as the ostriches, have downy plumes, in which many of the essential characters of the feather are greatly obscured. In the emu, whose habits are more strictly cursorial, the plumage almost resembles hair. In the cassowary the likeness becomes yet more striking, while the wingless apteryx of New Zealand has not even the few bare quills which stand for wing-feathers in the former bird. So, too, among those sedentary marine birds, the penguins, where the wings have been converted into a sort of fins for diving, the feathers undergo a parallel change into scales. There is reason, indeed, to suspect, as Mr. Lowne has pointed out, that these marine species retain in many ways the primitive characters of the class; and we may perhaps regard them rather as birds in whom the pinions and plumage have never fully developed than as birds in whom they have assumed a new form.

On the other hand, the truest feathers — that is to say, those which exhibit the essential features of a feather in the most marked manner — are specially connected with the act of flight. The general surface of the body is covered with soft down, among which sprout the delicate plumes that form the common covering for warmth and protection; but only on the wings and tail do those long and stiff quills appear which, after all, are the feathers par excellence, the models and prototypes of all the rest. Now, it is quite obvious to everyone that the wings are the organs of flight, and that the quills are the part by means of which the powerful muscles of the bird are brought to bear upon the sustaining atmosphere. As for the tail, its functions resemble those of a rudder, in directing the course of flight to right or left. The difference between these true flying feathers and the mere clothing of the back and breast is so striking that naturalists have given them separate technical names, as quills and plumes respectively.

From such facts, and others like them, I think we may arrive at an important conclusion — that feathers have been developed and selected through the habit of flight. Probably our monstrous friend the pterodactyl had only a membranous wing or bit of skin, extending from the elongated outer finger of his forearm to the leg. Such a parachute we still see among the so-called flying-squirrels and lemurs; while in the bats it has developed into a sort of webbed wing. But if any of the early birds happened to possess an altered hair-like or scale-like covering — the relic, perhaps, of some common reptilio-mammalian ancestor — which afforded them any extra grip upon the air through which they fell rather than floated, then those individuals would thereby gain an extra chance of catching prey or escaping enemies, and therefore of survival in the constant rivalry of species with species. The more perfect these organs became, the more closely adapted to the function of flight, the greater the advantage the bird would derive from their possession, and therefore the better the chance of survival which it would obtain. Thus, apparently, the most aërial birds have the largest and strongest quills, and the most quill-like plumes, while the running and diving birds have either never developed these adjuncts in their highest form, or else have lost them by disuse.

Let me take down one of the peacock's feathers, which stands on the mantelpiece in this Vallauris vase, and closely examine its structure. It consists of a long central shaft, horny and tubular at the lower end, and filled above with a soft, white, spongy matter; while a number of little barbed branches are given off on either side, curiously interlaced by means of tiny hooked filaments, whose myriad threads are far too numerous for the most industrious critic to count up. Everybody knows that this tubular structure combines in the highest degree the mechanical requisites of lightness and strength; and everybody has read that it is employed with the self-same object by human engineers, in such constructions as the great bridges which span the Menai Straits or the St. Lawrence at Montreal. Evidently this peacock's feather, though now converted to a purely ornamental function, was originally developed for the purpose of flight. If I doubt it for a moment, I need only look at the quill-pen in my desk over yonder. That flat blade, close-textured and strongly woven, clearly belongs to a flying organ; and this beautiful mass of green and golden waving plumelets is evidently modeled on the self-same plan. It is useless, or next to useless, now, for flight; but it still bears clear traces of its original function in the structure and arrangement of its shaft and barbs.

Next, let me look at the little downy feather I have abstracted from the Indian cushion. This is not a flying organ, nor did its representative on any early ancestor ever fulfill a similar office. Light, warm, soft, fluffy, its whole object is decidedly that of clothing against chilly weather, and protection against thorns or other rough bodies. Yet when I examine it closely, I see that the same general ground-plan still runs through it, as that which ran through the goose-quill and the peacock's tail-covert. "How comes this?" I ask myself; "here we have a small, delicate, almost fleshy shaft, instead of the horny quill; and a feeble set of downy barbs instead of the strong, well-woven blade: yet the main features remain unaltered, though the function is entirely different. How can I account for this resemblance?"

The case of the emu and the apteryx helps to throw light upon the problem thus disclosed. Where birds fly very little, their feathers never acquire or else soon lose the distinctive quill-like character; but where birds fly much, the quill-producing tendency becomes strong and pronounced. Primarily, this tendency ought to affect only those parts which are used in flight, namely, the wings and tail; and, as a matter of fact, we have seen that these are the parts which exhibit it in the highest degree. It would be almost impossible, however, that a change of such magnitude should be set up in some of the feathers, without to a lesser extent affecting all the rest. We might as well expect that the hair on a certain patch of some animal's skin would grow thick and spike-like, without any corresponding alteration in the rest of his body. True, natural selection does sometimes produce this result for some special purpose, when it is highly desirable that an acquired character should be confined to a small area. But, as a rule, when one part of the skin hardens, like that of a turtle or crocodile, the tendency to bony development shows itself in every part; and when certain hairs become converted into thick spines, like those of the hedgehog, the echidna, and the porcupine, a general bristly tone pervades almost all the coat. The scaly plates of the armadillo and the pangolin in like manner communicate a universal scaliness to the whole external surface of the animal. We may say in simple language that the body has got into the habit of producing certain structures, and that the habit extends to analogous parts in which it is not strictly necessary.

This is the case with the flying birds. Some of their feathers — modified scales or hairs — having become specially adapted for flying, all the rest follow suit to a greater or less extent. Indeed, we can hardly imagine how quills could come into existence at all, unless we allow that there must first have been an adventitious tendency toward the production of light-barbed shafts over the whole body. Those birds which exhibited this adventitious habit in the highest degree would become the ancestors of the aërial species, in whom it is still further developed by natural selection; while those birds which exhibited it in the least degree would become the ancestors of the diving, running, and scraping tribes, in whom natural selection favors rather such special adaptations as web-feet, fin-like wings, long and powerful legs, and ornamental plumage.[1]

The æsthetic philosopher, however (if the reader will permit me to designate myself by such a periphrasis), is far more interested in the modifications which feathers undergo, after they have become feathers, than in those which they undergo before reaching that stage of their development. For the infinite variety of coloring, the exquisite tones of metallic sheen, the graceful arrangements of crests, tufts, plumes, and lappets, which render birds such conspicuous objects in our museums or gardens, are all of them due to the pigments or shapes of feathers, and all of them have apparently been produced by the voluntary choice of beautiful mates among the birds themselves.

The modifications of feathers thus originated form, of course, a clew to the tastes of the various birds which possess them; because each species will naturally select such mates as best satisfy its ideas of the beautiful, and so will transmit the admired qualities to its descendants. It is a remarkable fact that the tastes of many birds, indirectly disclosed in such a manner, coincide very closely with the tastes of mankind at large.

Not all birds, however, exhibit equally these æsthetic preferences. Some large families, like those of the hawks, eagles, owls, and nightjars, are noticeable neither for beauty of color nor for richness of song. Other classes, again, like those of our own English hedge-birds, seem rather musical than chromatically inclined in their tastes. As a rule, we may say that birds of prey and nocturnal birds are very deficient in aesthetic feeling, all their energies being apparently directed to swiftness of pursuit and skill in hunting; while, on the other hand, small seed-eating birds, and those which live on little insects or other minute animals, generally expend all their æsthetic sentiment on the faculty of song. But only those birds which live upon fruits, or the mixed nectar and insects extracted from flowers, usually possess brilliant colors.

I have already more than once pointed out to the readers of the "Cornhill Magazine" the probable reason for this peculiar connection.[2] The eyes of fruit-eating or flower-feeding animals become specially adapted to the stimulation of colored light, and therefore the creatures become capable of receiving special pleasure from such sources. Accordingly, those among their fellows which displayed brilliant colors would prove most attractive, and would be chosen as mates for their beauty. I have instanced before, among the flower-feeding species, the numberless varieties of humming-birds, and the almost equal profusion of sun-birds, to which we may add a few other minor forms, such as the brush-tongued lories; while among the fruit-eaters, the parrots, macaws, cockatoos, toucans, barbets, nutmeg-pigeons, fruit-pigeons, chatterers, and birds-of-paradise, may stand as cases in point. But it will be more interesting here to glance briefly at the various modes in which these colors are produced than to extend the list of species which display them. The commonest method of exhibiting color is by means of pigments either in the external coating of the feathers or in their deeper layers. Cases of this sort are too frequent to need special exemplification; but some birds have brilliant hues otherwise displayed, as in the wattles of the common barn-door fowl, the fleshy appendages of the turkey, and the painted face of the carrier-pigeon. The wattled honey-sucker of Australia has two drooping folds of flesh which fall like bonnet-strings under his throat; the king-vulture has his head and neck covered with naked skin of every hue in the rainbow; and the cassowary (by far the most frugivorous of all the ostrich tribe) has the same parts of a brilliant red, variegated with melting shades of blue. In many other birds the beak becomes an ornamental adjunct; and this tendency reaches its furthest development in the bill of the toucan, whose colors almost vie with the humming-bird itself. But the most curious of all such aesthetic modifications is that from which the wax-wings derive their name. In these birds the shafts of certain wing-feathers are prolonged into small, horny expansions, bright scarlet in hue, exactly resembling, both in color and texture, little tags of red sealing-wax.

The metallic luster of feathers is generally due to fine lines on the surface of the barbules, like those which produce the iridescence of mother-of-pearl. Such luster occurs in the sun-birds and hummingbirds, and on many other less ornamented species. Sometimes gleaming like gold or bronze, sometimes fading away into jetty black, anon reappearing as glancing outbursts of crimson, azure, or exquisite green, it has gained for the birds on which it appears such poetical names as ruby-throated, topaz-crested, amethystine, golden, emerald, and sapphire. Not only does it occur upon the burnished neck of the dove, but it gives a passing splendor to the sable livery of the crow, and throws a thousand changeful hues over the glossy plumage of the mallard.

But besides the ornamental effects of color and luster, feathers appeal to the æsthetic taste of birds by their form, their arrangement, and their variety. Only the plainest birds have all their plumage exactly uniform and simply disposed. In an immense number of species certain feathers have been specially modified in shape so as to form crests, fan-like tails, lappets, and other ornaments. And just as a good architect lavishes his decorations chiefly on the constructive points of his building, the critical parts, such as arches, doorways, windows, and architraves, so do we find that birds have chosen to place their decorative modifications on the most important nodal points of their bodies, and that they generally lavish their richest coloring upon these ornamental adjuncts. This peacock's feather, for instance, formed part of a gorgeous semicircular fan, which composed, as it were, the background or reredos of the whole living picture when expanded, and the train of the majestic sultan when folded in repose. A plume from the neck or back, though still beautiful with golden green and faintly purplish blue, would not have exhibited those splendid eye-like spots which reflect the sunlight in a mingled mass of glory from this perfect tail-covert. Only in the most fitting positions for decoration do birds, as a rule, expend their choicest designs.[3]

The feathers of the ostrich naturally occur first to the human investigator of æsthetic taste in birds. The quills of the wing and tail, here purely ornamental in their function, compose the well-known silky plumes of commerce. The common crane has also beautiful elongated wing-feathers, which fall on either side of the tail in graceful waving masses. If we may trust the doubtful pictures which have come down to us, that grotesque and gigantic pigeon, the dodo, possessed similar tufts of ornamental plumage. But the great order of gallinaceous birds, or the hen and turkey tribe, display the most magnificent tails of all, so familiarly known in the peacock and the pheasant family, as well as in the humbler denizens of our English farmyards.

Crests form another favorite ornamental device among birds, occurring independently in the most different orders. The graceful tuft of the gray heron must have attracted the attention of every observer. Among the pheasants similar decorative adjuncts are common; and the curassow shows this peculiarity in a very beautiful form. With parrots and cockatoos, crests are of frequent occurrence, and they make equally striking features among the humming-birds and sun-birds. Indeed, it may be roughly asserted that those birds which seek their food among flowers and fruits, and which consequently exhibit a taste for bright colors, are also the species in which ornamental tufts of feathers most frequently occur. But crests are also found even among the generally somber and inartistic birds of prey, being by no means unusual in the owls and hawks, while the serpent-eating secretary-bird derives his queer name from the fancied resemblance of his top-knot to a pen stuck behind the ear. Other well-known instances of crested species are the hoopoe, the wax-wing, the golden-crested wren, and many jays. But the umbrella-bird, a Brazilian fruit-crow, exhibits the fullest development of this particular ornament, having the whole head covered by a dome of slender, shining blue feathers, about five inches in length by four and a half in breadth. It may be added that almost all birds which possess these ornaments possess also the power of raising or depressing them at will; and that during the season of courtship the male birds constantly expand all their charms before the eyes of their admiring mates. We have all seen this ostentatious display ourselves in the case of the peacock, the turkey, and the barn-door fowl. It proves almost beyond a doubt the aesthetic purpose and function of such otherwise useless, inconvenient, arid vitally expensive excrescences.

Sometimes the crest is produced by some other means than that of a mass of plumes. Besides the well-known fleshy comb of our friend chanticleer, there is the horny helmet of our old acquaintance the cassowary, and the quaint protuberances on the beak of the jacana. Most eccentric of all is the device adopted by the hornbills, whose name sufficiently indicates their peculiarity in this respect. The beak in these birds is prolonged above into a single unicorn-like process, extravagantly disproportioned to the general size of its wearer.

On the other hand, it may be noted that most small singing-birds, or other species which live on seeds, grains, insects, and mixed small food, are destitute of tufted ornaments, as well as of brilliant coloring.

The lappets, frills, or other neck-pieces of so many decorated species must not pass entirely unnoticed in this review of æsthetic devices among birds. Beginning with the mere burnished breast-plumage of the pigeon, or the crimson stomacher of the robin, they become at last, in the humming-birds, sun-birds, and other tropical species, the most exquisite drapery of amethyst, topaz, emerald, or golden bronze. The so-called beard of the turkey is a special example of a very aberrant type. The ruff derives his English name from a similar peculiarity.

The birds-of-paradise unite all these modes of ornamentation in the highest degree, and with the most harmonious results. They join the graceful plumes of the ostrich to the dainty coloring of the sun-bird. Crests almost as largely developed as that of the umbrella-bird overshadow their beautiful heads; frills as full as those of the hummingbirds fall down in metallic splendor before their gorgeous necks. And, if any proof be wanting of the connection between the nature-of the food and the general beauty of the plumage, it may be found in the fact that these royally-attired creatures are first cousins of our own dingy crows and jackdaws; but, while the crow seeks his livelihood among the insects and carrion of an English plowed field, the bird-of-paradise regales his lordly palate on the crimson and purple fruits which gleam out amid the embowering foliage of Malayan forests.

Equally magnificent are the members of the genus Epimachus, inhabitants of the same brilliant archipelago. Their long, silky plumes float behind them in the same graceful curves; their burnished necks are adorned with the same glancing hues of ruby and emerald. Yet they are surpassed in one respect by their distant relatives, the lyrebirds, first cousins of our diminutive English wrens. Though destitute of brilliant coloring and metallic sheen, these curious birds exhibit in their long and beautiful tails the only undoubted example among the lower animals of a love for symmetrical patterns.

I have only bethought me now of a few among the countless modifications which feathers undergo, for the aesthetic gratification of their wearers, or rather of their wearers' mates, and the list might be almost indefinitely prolonged. But it will he better worth while, perhaps, to glance briefly at another set of facts connected with feathers — I mean their artificial employment by human beings for the exactly identical purpose of aesthetic decoration. Could any fact show more clearly the similarity of artistic feeling which runs through the whole animal series than this thought, that man makes use, for his own adornment, of the very self-same beautiful colored baubles which the birds originally developed to charm the eyes of their fastidious brides?

I need not recall by name the various kinds of plumage so employed — the feathers of the ostrich, the marabou, the bird-of-paradise, the emu, the pheasant, and the gull; the sun-birds and the hummingbirds mercilessly slaughtered by the million in the Malay Archipelago, Ceylon, and Trinidad to supply the bonnets of London and Paris; the swan's-down, the grebe, the widow-birds, the cockatoos, the parrots, the macaws, which decorate our wives and children with barbaric spoils. It will suffice to remember, in passing, that from the feather mantles of Hawaian kings, the feather kirtles of American Indians, and the feather mosaics of Mexico, to the plumes of our own court-dress, our own military uniforms, and our own quaintly surviving funeral processions, these same "dermal modifications" of birds have served an aesthetic purpose, better or worse, throughout the whole course of human history.

Nor does the resemblance stop here. Mankind employs tufts of feathers for decorative display in just the same manner as the birds who originally developed them. The Red Indian in his war-paint dressed out his head with a row of quills, arranged in exactly the same order as the top-knot of a hoopoe or a cockatoo. The feather collars of so many savage tribes recall to the letter the frills and lappets of the humming-bird or the epimachus. The ostrich-plumes of our English royal receptions, and the panache of our European officers' dress, are adaptations from the primitive idea of the crane and the umbrella-bird. Everywhere, the tuft of feathers is placed on some prominent part of the person — some "constructive point" in the human or avian system of architecture.

A ring at the bell warns me that a visitor is standing at the door. I throw my little feather hastily into the fire, and cut short my reflections to welcome my expected guest. But one last thought occurs to me before I close my afternoon's meditation. To be "pleased with a feather" appeared to the great metaphysical poet of the eighteenth century a mark of childish simplicity. Perhaps it may be so; but, after all, is there not some solace in that new philosophy which can enable one to pass a whole hour, this murky afternoon, in pleasurable contemplation of that tiny plume which seems no contemptible subject of human study to Charles Darwin and Herbert Spencer?

1 - Of course no effect, in nature is really accidental, that is to say, uncaused; but, in organic nature, effects which arise from special collocations of causes, unconnected with the previous habits of a plant or animal, may fairly be called adventitious. If they result in some alteration beneficial to the species, the alteration will be further strengthened by natural selection, and its final outcome will be a purposive structure — that is to say, a structure specially adapted to its peculiar function. But it must be remembered that almost all purposive structures were in their origin adventitious. I say "almost all" and not "all," because an exception must be made in favor of what Mr. Herbert Spencer calls "functionally-produced structures."

2 See a paper on "The Origin of Flowers," in "The Popular Science Monthly Supplement" for June, 1878; and another on "The Origin of Fruits," in "The Popular Science Monthly" for September, 1878.

3 - I say "as a rule," because the hornbills, toucans, vultures, certain pigeons, and a few other species, offend against our ordinary human canons of taste; but the ornaments of birds seldom or never render them ridiculous in our eyes, like those of many highly decorated monkeys.

GEOLOGY AND HISTORY

The science of human life has been the last to recognize that minute interaction of all the sciences which every other department of knowledge now readily admits. We allow at once that no man can be a good physiologist unless he possesses a previous acquaintance with anatomy and chemistry. The chemist, in turn, must know something of physics, while the physicist can not move a step until he calls in the mathematician to his aid. Astronomy long appeared to be an isolated study, requiring nothing more than geometrical and arithmetical skills but spectrum analysis has lately shown us its intimate interdependence upon chemistry and experimental physics. Thus the whole circle of the sciences has become a continuous chain of cycles and epicycles, rather than a simple sequence of unconnected and independent principles.

History, however, still stands to a great extent outside the ever-widening sphere of physical philosophy. It is comparatively seldom that we see an historian like Dr. Curtius acknowledging the interaction of land and people upon one another's character and destiny. More often we find even the modern annalist writing in the spirit of Mr. Freeman, as though men and women formed the only factors in the historical problem, and the great physical powers of Nature counted for nothing in the game of human life. Yet a few simple instances will show at once the fallacy of such a view. If the ancestors of the Hellenic people had gone to the central plains of Russia instead of to the island-studded waters of the Ægean, could they ever have produced the magnificent Hellenic nationality with which we are familiar? Was not their navigation the direct result of their geographical position on the shores of an inland sea, intersected by jutting peninsulas, and bridged over by a constant succession of islands, each within full sight of its nearest neighbors? Was not their polity predetermined in large measure by the shape of their little mountain valleys, each open to the seaward in front and closed by a natural barrier of hills in the rear? Could their plastic genius have risen to the height of the Olympian Zeus and the Athene of Phidias if they had possessed no material for sculpture more tractable than the hard granite of Syene? While we allow that the Aryan blood of the Hellenes had much to do with the differences which mark them off from the Negroid Egyptians, can we doubt that Hellenic civilization would have been very different if the settlers of Attica had happened rather to occupy the valley of the Nile; and that the Egyptians would have become a race of enterprising sailors and foreign merchants if they had chanced to make their homes on the shores of the Cyclades and the Corinthian Gulf?

Or, again, let us look for a moment at Britain. Who can suppose that the destiny of our country has not been profoundly affected by the existence of great coal-fields beneath its surface? Even if we possessed no mineral wealth, it is probable that our geographical position would still have insured us a considerable commercial importance as the carriers of the civilized world. Britain happens to occupy the central point in the hemisphere of greatest land, and this fact, aided by its insular nature, could not fail to make it a great mercantile country as soon as navigation, nursed in the Mediterranean, had advanced sufficiently to embrace the whole ocean-coasts of Asia, Africa, and

America. But without coal and iron we should have been mere merchants, not manufacturers. London, Liverpool, Glasgow, and Southampton might possibly have been not inconsiderable marts for exchanging the products of other countries, and for balancing the trade in raw cotton or sugar from India and America against the textile fabrics and the hardware of France and Belgium. But we should have had no Birmingham, no Manchester, no Sheffield, no Leeds, no Bradford, no Paisley, no Belfast. Our population would not have reached one half its present size. Lancashire, the West Riding of Yorkshire, and the busy mining district of South Wales would be as thinly inhabited as Merionethshire and Connemara. The Black Country would be a quiet pastoral and agricultural region like the remainder of Warwick and Stafford. We should have no great towns except on the seaboard and the navigable rivers, and even these would only attain a fraction of their existing dimensions. Most of our people would be engaged in farming, and there would be no great wealthy class to crowd into Brighton, Scarborough, Cheltenham, Torquay, and the Scottish Highlands.

But this is not all: the difference in our national character would no doubt be very great. Coal has stimulated our inventive faculties and our enterprise, and has given an indirect impetus to science and art. Without it we should have had fewer mechanical improvements, fewer scientific discoveries, fewer railways, fewer colleges and schools. All these things have reacted upon our general level of intelligence and taste, and have enabled us to hold our own among the most advanced European nations. But without coal and iron we should have fallen back to somewhat the same position as that now held by Holland or Scandinavia, allowance being made for a larger territory in the first case, and a thicker population in the second. Our comparatively insignificant numbers would reduce us from the rank of a first-class European power to that of a nation existing on sufferance. Our army and navy would be smaller; our Parliament less important and less stimulating to high ambitions; our churches, our bar, our medical faculty less advanced in the fore-front of thought. Thus we should probably suffer in every respect, producing both absolutely and relatively fewer great men, either as thinkers, administrators, discoverers, inventors, or artists. For, when once a nation has fallen behind in the race, the audience addressed becomes smaller, the competition less keen as an incentive to effort, the rewards of success decrease in value, and the general atmosphere of example and rivalry deteriorates in power. Where few books are written, few investigations undertaken, few works of art produced, few and still fewer care to aspire toward a forgotten ideal. Thus, without coal, Britain might have declined from the England of Shakespeare, Milton, and Newton, just as other countries have ^declined from the Hellas of Pericles and Plato, and the Spain of Cervantes and Velasquez.

The relation between physical conditions and history in its wider acceptation being thus fundamental, it may be well to consider in somewhat greater detail the special reactions of a single tolerably definite portion of the natural environment upon human development. For this purpose we may choose the science of geology. It might seem at first sight that geological facts had very little to do with the course of history. Rocks and clays, lying often far beneath the surface, and comparatively disregarded till a late stage of civilization, would appear far less important in the evolution of mankind than plants and animals, geographical situation and meteorological conditions. But, though doubtless of inferior practical interest to these superficial phenomena, the geological constitution of the soil is yet pregnant with innumerable reactions upon the life of human beings who dwell upon its surface. I hope to show in the sequel that the rocks or minerals which lie beneath the thin coating of earth and vegetation have always exerted an immense though often unsuspected influence upon the history of man. And I shall choose most of my examples from well-known facts of the British Isles, only diverging elsewhere very occasionally for the sake of more striking or more conclusive instances.

To begin with, it must be premised that geological conditions were of comparatively less importance in very primitive times, and have increased in their practical relation to humanity with every additional step in general culture. This is only what we must expect from the nature of the case. Man's connection with his environment has necessarily grown more and more complex as his evolution proceeded. Soil becomes a matter of interest sooner than building-stone; potter's clay precedes copper or iron ore as a valuable object; metals of every kind are earlier required than coal. The mere savage needs nothing more from the mineral world than flint for his arrow-head and ochre for his personal adornment. A little later he requires bronze for his hatchet, gold and amber for his rude jewelry, clay for his hand-molded earthenware. A still more advanced race will learn to prize silver for coins, lapis lazuli for gems, brick-earth for Assyrian temples, granite for Egyptian colossi, marble for Hellenic sculpture, and iron for Roman swords. Only at a very late period of development will man begin to be largely affected by the neighborhood of zinc, lead, and mercury, of rock-salt, kaolin, and plumbago, of slate-quarries, marl-pits, and pipe clay beds. Last of all will come the economic employment of coal, which in our own island has caused the aggregation of densely massed populations around the great centers of Glasgow, Manchester, Leeds, Sheffield. Newcastle, and Birmingham.

How general is the relation in early stages of civilization we can see from the comparatively close similarity between the life and arts of all the lowest savages. How special it becomes in advanced societies we can see when we consider the cases of Bethesda growing up by the side of the Penrhyn slate-quarries; of Broseley, entirely engaged in the manufacture of clay tobacco-pipes; and of Northwich, Middlewich, and Nantwich supporting themselves by mining rock-salt.

Nevertheless, even at the earliest period, geological conditions must have largely influenced human life. Tribes which lived among rugged granite or limestone mountains must have been very differently circumstanced from those which ranged over level tertiary lowlands, or settled on the alluvial deltas of modern rivers. During that primitive epoch which Sir John Lubbock has christened the Palæolithic age, when man first dwelt in Britain, we see traces of such primeval differentiation. The naked or skin-clad savages, who then hid among the caves of southeastern England, were ignorant of all the metals, as well as of pottery, and only employed rudely chipped weapons of unground flint. The neighboring forests then contained the mammoth and the woolly rhinoceros, the urus and the musk-ox, while the hippopotamus still basked on the banks of the Ouse and the Thames. But man appears at that period to have been wholly confined to the southeastern corner of England, from the coast of Devonshire to that of Lincoln. This district roughly coincides with that in which he could obtain flints for the manufacture of his weapons; and it also comprises the most level portion of Britain, where he might find comparative security and well-stocked hunting-grounds among the low-lying jungles of the eastern counties, the Thames Valley, and the tertiary plains of Hampshire. He does not seem at this early age to have ventured among the wild primary hills of Cornwall, Wales, the Pennine chain, and the Scottish Highlands, but rather to have clung about the river fisheries and the flat shores of the southeast. Perhaps the bare and treeless chalk downs whicH run from Beer in Devonshire to the Norfolk coast, backed by a forest-belt on the oölite in the rear, may have checked his westward advance through the fear of meeting the cave-lions and other savage wild beasts of the preglacial period on the open plain.

At a far later date, when man had progressed from the hunting to the pastoral stage, and had learned to fashion weapons of polished stone or bronze, which made him the acknowledged master of the brute creation, it is clear that a great change must have taken place as regards the relation to geological conditions. And in Britain the men of this later period certainly spread over the whole country, gathering most thickly, it would seem, where pasturage was easiest for their herds and flocks. This would naturally be upon those same undulating chalk downs which were doubtless objects of terror to the earlier race. Hence we find the tumuli and other memorials of the Euskarian and Keltic inhabitants—belonging either to the neolithic, the bronze, or the iron age—most thickly clustered around the great monument of Stonehenge on Salisbury Plain, among the downs about

Brighton or Lewes, and on the sides or summits of the Yorkshire and Lincolnshire wolds. In those days and for many centuries after, the Weald of Kent lay as a wild forest-belt between the open chalk country to the north and south; while the primary hills and the river valleys still consisted for the most part of unbroken underbrush and woodland. Even in these early times, however, a commerce based upon geological differences had already sprung up: for the beautiful jade, employed as material for the finest hatchets, has been recognized as coming from the Kuen-Lun Mountains of Central Asia, while amber was already imported from the banks of the Baltic. Within Britain itself the Cornish tin-mines probably supplied the metal which mingled with copper to form the bronze implements of all Western Europe. An industrial population must even then have gathered with comparatively considerable density above the ores of the Land's End, while the valley of the Thames remained a mere desolate jungle wandered over by a few stray families of savage hunters.

Agriculture must first have developed itself over the whole world on low alluvial ground. Hence we find that all the great early civilizations occupy river valleys—such as those of the Nile, the Euphrates, the Ganges, the Indus, and the Hoang-Ho. Here alone can large masses of men obtain subsistence, before navigation and scientific agriculture have reached a considerable stage of evolution. Here, too, the density of the population and the level nature of the soil permit the growth of those vast despotisms under which alone an early society can be organized with any high degree of internal diversity. But just as navigation, nursed on inland and island-studded seas, spreads afterward to the wider oceans, so agriculture, nursed on well watered alluvial plains, spreads afterward to drier, rockier, or more mountainous districts. In the desert uplands of the Punjaub, cultivation exists wherever wells can be sunk, even at immense depths, and the industrious Jat peasantry work ceaselessly day and night by relays, each family raising the precious water to fertilize its own little plot, for a stated number of hours out of the twenty-four. But such industry presupposes a long training in more fertile soils, and a heavy pressure of population on all the earlier occupied alluvial lowlands. So too in Britain, a primitive agriculture would have despaired of raising corn upon the bare sides of the Chiltern Hills, and only modern scientific farming has turned the boggy upland expanses of the Cheviots and Lammermoor into nourishing tillage. Accordingly, we might expect that the growth of agriculture would bring geology and human development into still closer connection within our island.

Geologically, Britain falls into two well-marked divisions—the northwestern primary tract, and the southeastern secondary and tertiary region. The boundary between them may be roughly marked off by a line running from the mouth of the Tees to the mouth of the Exe. Northwestward of this line we have the whole of Scotland, the Pennine region of England, the Welsh mountain system, and the peninsula of Devonshire and Cornwall. Southeastward we have the whole level country of England, comprising the plain of York, the great central plateau, the Fen district and the eastern counties, the valley of the Thames, and the watershed of the south coast.

Now, it is not too much to say, that by far the most fundamental fact in the annals of Britain, since the dawn of written history, is the great revolution which has exactly reversed the relative importance of these two divisions. Yet what are called histories of England at the present day utterly ignore that revolution. In the Roman period and the middle ages, the most valuable and most populous part of Britain was the secondary and tertiary lowland: at the present day, the most valuable and most populous part is the primary division to the north and west. And what gives this revolution its greatest ethnological interest is the fact that while the secondary tract roughly corresponds with the Teutonic portion of Britain, the primary tract roughly corresponds with the Keltic or semi-Keltic portion.

As early as the time when Caius Cæsar, the Dictator, landed in our island, these two great divisions had already shown their differentiating characteristics. The Britons of the southeastern country, consisting of open and easily cultivable plains, had advanced to the agricultural stage, and were comparatively dense in their pressure upon soil, with fixed habitations and considerable towns. The northwestern tribes were still pastoral nomads or hunters, dwelling in movable villages, and having mere empty forts on the hill-tops, to which the whole population retreated in case of invasion. The difference thus expressed continued more or less marked throughout the whole historical period, until the use of coal effected that extraordinary revolution by which primary and industrial Britain has at length asserted its superiority to the level agricultural southeast.

Under the Romans Britain became a corn-producing and grain exporting agricultural country, like the America of our own day. And just as the valley of the St. Lawrence and the northern Mississippi basin now form the most important wheat-growing part of America, so the valleys of the greater rivers formed the most important part of Roman Britain. The plain of York, formed by the Ouse and other tributaries of the Humber, is the largest low-lying corn-field and meadow-land in our country. It consists mainly of triassic strata, overlaid in the lower reaches by a deep bed of alluvium. In the center of this rich agricultural tract lay the Roman provincial capital of Eboracum. Another wealthy region is the post-tertiary level of the eastern counties; and here the colony of Camalodunum lay surrounded by numerous villas of rich land-owners. The tertiary valley of the Thames shows its importance by including the considerable cities of Londinium, Verulamium, and Rhutupite. Other Roman towns—Lincoln, Cirencester, Bath, and Dorchester—filled up the rich oölitic and green-sand belt of central England; while Winchester overlooked the tertiary vale of the Itchin at Southampton, and took its name of Venta Belgarum from the-agricultural lowland at its doors. We may gather from the Roman historians that the occupation of southeastern Britain was real and thorough. The native population was reduced to serfdom, and the country became a mere feeder of Rome or of the Gallic cities.

Primary Britain, however, seems never to have fallen into so miserable a condition. The Roman supremacy was here probably confined to a mere military occupation, like our own occupation of Kumaon or the Simla Hills. Caledonia never fell into their hands, and even in Wales and the Pennine chain we find only military stations, like Isca Silurum or Segontium, not large cities like London, York, and Lincoln. Even where the Romans thoroughly penetrated the primary region, as in Cornwall or the Forest of Dean, it was always for a geological reason, to secure the mines of tin or iron. This difference, I believe, had almost as much to do as geographical position with the subsequent relations of the Britons to the English invaders. While the servile herd of the Belgian, Icenian, Trinobantian, and Brigantian country, demoralized by Roman centralization, fell easily before the Jutish or Anglian pirates, the more independent mountaineers of Wales, Cumbria, and Strathclyde long resisted the English onslaught, and only at last succumbed as free subject races, instead of being enslaved or exterminated like their eastern fellow countrymen. The Scottish Highlands not only retained their own independence, but even gave their kings to the Teutonic Lothians. Granite naturally makes freemen, as alluvium naturally makes slaves.

When the English settled in southeastern Britain, they occupied for the most part the secondary and tertiary plain. But they also pushed northward into the primary region up to the Firth of Forth, as the Romans had done before them. The Teutonic invaders, in other words, took the best agricultural lands for themselves, while the Kelts were driven back into the rugged primary tract of hill and forest. Throughout the middle ages, agriculture and grazing formed the staple English industries. Accordingly, during the early English period, we find all the more important towns occupying the cultivable valleys or gentle plains. Canterbury and Rochester, the two Kentish capitals, stand in the midst of tertiary lowlands; London, the final royal city of the West Saxon kingdom, lies surrounded by a similar tract; the Oxfordshire Dorchester, first home of the Wessex kings, is on the border of the rich vales of Aylesbury and Oxford; Winchester, their later seat, commands the valleys of the Itchin and the Test. Norwich, Bury St. Edmunds, and Ipswich were important centers for the East Anglian drift. Peterborough and Ely rose among the levels of the Nen and the fens of the Ouse. Lincoln, Oxford, and Chippenham stood upon the great central oölitic belt. Cambridge occupied a low-lying corner of the cretaceous system. Exeter, Lichfield, and Chester were girt round with the fertile triassic meadow-lands. York still remained the capital of the north, and the metropolis of a kingdom which long retained the foremost position held by the north under Roman rule. These were the great cities of England before the Norman Conquest, and not one of them stands upon a primary formation. All of them, save only London, have now sunk to the position of mere cathedral cities, university towns, or agricultural centers. But Edinburgh, Glasgow, Manchester, Leeds, Sheffield, Newcastle, Bristol, and Cardiff, the great cities of to-day, are all built upon primary rocks; while the only two important modern towns which rest on later strata are Birmingham, on the borders of the Black Country coal-field, and Liverpool, which lives by conveying the cotton of America to the great Lancashire colliery district around Manchester, Rochdale, and Oldham.

In the later middle ages England became a wool-stapling country. Bales of wool were shipped from the Orwell for Flanders and Italy, as they are now shipped from Australia for Leeds and Bradford. This was the first step toward making Britain a commercial country. Before the Norman Conquest it had been an essentially agricultural and self-sufficing community, growing all that it required to meet its own simple needs, and neither exporting nor importing goods to any noticeable extent. But the wool export created a foreign trade. Ports sprang up along the south and east coasts, from Dartmouth, Topsham, and Lyme Regis to the now forgotten haven of Ravenspuron-Humber, the precursor of our modern Hull. This trade gave importance to the chalk districts, high sheep-walks now the barest and least inhabited portion of southeastern England. Not a single town of any pretensions at present occurs in any part of the downs or wolds. But Dorchester, Shaftesbury, Old and New Sarum, Winchester, Lewes, Reading, Wallingford, Cambridge, and Beverley, were all places of great mediæval importance, and all stand within the cretaceous area. Other wool-growing tracts of course possessed a similar value.

A few more special agricultural features of the various secondary or tertiary geological formations may here he fitly introduced. The Trias and other "Poikilitic" strata, running across England from the Tyne to the Exe, form beautiful undulating country, comprising much of the best wheat-growing and pasture land, and famous for the production of cheese. In this belt lie the vale of York, the Trent and Severn Valleys, the Cheshire Plain, and the vales of Exeter and Taunton. An outlier forms the valley of the Eden at Carlisle. The Lias, which follows the Poikilitic series to the southeast, is a good soil for corn and apples, but also produces the most excellent cheese in England, as Mr. Woodward has pointed out. Along the Severn bank it furnishes the double Gloucester; at Melton Mowbray and Leicester it produces Stilton; and in Somersetshire it unites with the triassic red marl to yield the Cheddar. The fruitful vales of Eversham and Gloucester belong to this formation. The Oölite gives us the rock known as cornbrash, which disintegrates into a splendid wheat-bearing soil, naturally manured by its large quantities of phosphate of lime, the so-called bone-earth. The Oxford Clay, on the other hand, is poor and hard to cultivate, so that most of it lies under permanent pasture. It forms the sheep-feeding vale of Blackmore, in Dorset. The Kimmeridge Clay, in like manner, does not repay cultivation, and is mostly employed for meadow or woodland. The Wealden, forming the great trough between the North and South Downs, is another of the infertile soils. It remained a great wood, the Andredesweald, or Forest of Anderida, for a long period after the English conquest, and the local names of the district still retain their forestine terminations of hurst, ley, den, or field. Even at the present day the Weald is damp and clayey land, little tilled, and either laid down in pasture or given over to furze and heath. The Gault makes good grazing-lands, and the Upper Greensand is in every respect a fertile formation. These two series yield the rich Vale of White Horse, through which the Great Western Railway runs between Swindon and Didcot, as well as the vale of Aylesbury, whose name has become synonymous with pure milk. The Chalk supplies us with South Down mutton, said to owe its excellence not so much to the pasture itself as to a small land-snail (Helix virgata) which the sheep devour in great numbers.[1] The London clay, though stiff, can be made to yield good crops. Drift forms the great East Anglian plain, while the Fen country, the Somersetshire levels, and Holderness consist mainly of alluvium. Thus we see that, little as the mediæval farmer suspected it, the distribution of his corn-fields and pasture-lands, his orchards and sheep-walks, nay, even of the royal forests and the barren heaths, was finally dependent upon underlying geological conditions.

Even in mediæval and agricultural England, however, certain particular spots acquired a special industrial character from the nature of the subjacent strata. The occurrence of fuller's earth in the Stroud Valley and near Bath and Bradford gave rise to the west country cloth-trade. Salt was pumped from several inland wells in the Trias at Droitwich in Worcestershire, at Northwich, Sandbach, Middlewich, and Nantwich in Cheshire, and at Shirleywich in Staffordshire. The bays in which sea-water had been evaporated to yield salt had been known as "wyches," and the same word was applied to the new wychhouses of the interior. Clay suitable for potteries was found in many places, and naturally produced a small trade. But mines were little worked, and building-stone, of which more must be said hereafter, formed almost the only other geological differentiating factor between various districts.

The change to the modern industrial distribution is far too large a subject to be treated otherwise than quite cursorily here; but a few traits of the change may perhaps be sketched with a rapid pen. In Britain mineral wealth is almost universally connected with the primary formations. Our coal more especially has formed the great central pivot upon which turns the whole manufacturing and commercial system of the country. As soon, therefore, as the use of steam began to revolutionize our industrial world, the primary tracts of England, Wales, and Scotland, rose to the highest importance. The population of Britain suddenly found itself turned back upon the Keltic and coal-bearing regions. A slight classification of the various great towns of modern Britain according to the coal-fields in which they stand, or on which they depend, will serve to show the vastness of the revolution.

In or around the Scottish coal-field stand Glasgow, Paisley, and Greenock. Above the Tyne colliery region are Newcastle, North Shields, and Durham, while close at hand lie Sunderland, Stockton, Darlington, Middlesborough, and the Cleveland iron district. The Lancashire field incloses Manchester, Blackburn, Wigan, Bolton, St. Helens, Burnley, Middleton, Oldham, Rochdale, and Ashton, with Liverpool for its port, and Preston and Macclesfield upon its outskirts. An outlier contains Stoke-upon-Trent and Newcastle-under-Lyne. The West Riding coal-field includes Leeds, Bradford, Wakefield, Barnsley, Sheffield, and Chesterfield, while Huddersfield, Nottingham, and Derby hang upon its border, and Hull supplies it with an eastward outlet. The Staffordshire tract comprises Wolverhampton, Bilston, Dudley, Wednesbury, and Walsall, with Birmingham for its real center. Other carboniferous deposits occur in Coalbrookdale, in the crowded South Wales district, and near Bristol. If all these are put together, it will be seen that they compose almost all the great foci of British life and manufactures at the present day.

On the other hand, what are the great towns in the secondary and tertiary southeastern tract? London, the main distributing center, preserved by its navigable river, and its official importance. Southampton, a convenient Indian and South American port. Plymouth and Portsmouth, two government naval stations. Chatham, an artificial creation for purposes of war. Scarborough, Brighton, Cheltenham, Bath, and half a dozen other lounges for the moneyed classes. All these ultimately depend for existence upon the wealth created elsewhere. Leicester is almost the only town in purely Teutonic England which now earns a good livelihood by industries unconnected with the sea or with warlike preparations. Turning to the north, Edinburgh survives by its traditional position as a metropolis and as the center of the Scottish Church, the Scottish law, and to some extent the Scottish aristocracy, as well as by its possession of a university and a great cultivated society. But Edinburgh itself stands on a primary site.

The specialties of the modern system are far too numerous to allow even of passing exemplification. Here coal, there iron, in other places lead or tin, forms the source of wealth and the determining cause of human aggregation. The potteries draw men to Staffordshire; finer clays produce the ware of Worcester, Lambeth, or Dunmore. Flags for paving are largely worked in North Wales. Lime from blue lias keeps alive more than one small seacoast town. Even gold is mined near Dolgelley in Merionethshire. Phosphate of lime is collected as mineral manure. Cutler's green-stone and beds of jasper are found among the Cambrian rocks. Millstones, hearthstones, and fire-clay are other useful economic products. Terra-cotta is made at Watcombe, near Torquay. Epsom salts are manufactured from magnesian limestone on the Tyne. Slates for roofing, plumbago, Cairngorm pebbles, afford occupation in other parts to quarrymen and lapidaries. Glass can only be made where flints are obtainable. Whitby derives a small fortune from alum, jet, and the sale of fossils. Guernsey lives largely by exporting its granite as road metal to London. Whetstones supply an industry to Whittle Hill, and slate-pencils to Shap in Cumberland. But perhaps the strangest trade of all is that of the gun-flints, still manufactured at Brandon and Norwich to supply the savages of Africa, whither all the old flint-locks of Europe were shipped on the invention of percussion caps.[2] The water-supply everywhere depends upon geological conditions. Even our pleasure resorts and watering-places owe their attractions to similar considerations, as we can see when we examine the igneous masses of the Scotch Highlands, which form the chief heights of the Grampians; or when we remember that the self-same Cambrian rocks recur in the loveliest part of North Wales and in the Westmoreland lake district. So too in Devonshire, the regular tourist tract from Ilfracombe to Lynton and Lynmouth lies through the wild Devonian strata, which, interspersed with granite, once more reappear on the other tourist coast-line from Torquay to Land's End. Those who admire Ramsgate and Margate, with their bare, treeless downs and white chalk-cliffs, may also content themselves with the similar scenery of Dover, Folkstone, Eastbourne, or Brighton; but a different type of mind will prefer the wooded vales of Hastings, where the Weald comes down with its pleasant broken country to the seashore.

One last word may be given to the influence of geology upon art. We can hardly deny that the whole æsthetic development of Egypt must have been largely affected by its alternation between solid granite and the mud of the Nile. So, too, the Parthenon and the Apollo must have owed much to the marble of Paros and Pentelicus. China has doubtless been greatly influenced by the presence of kaolin clay. In Assyria, brick necessarily formed the chief building material; and in Upper India the monasteries and stupas of the Buddhist Emperor Asoka are still recognized by their huge, sun-dried bricks. Chryselephantine art could never alone produce high results; marble and alabaster would naturally yield far more elevated works. In Britain we may look for similar effects of the geological environment.

As early as the age when Stonehenge was piled up, building-stone was selected for special purposes, since the outer circle of that prehistoric monument consists of the Sarcen bowlders of the neighboring plain; but the inner pillars are of diabase, and have been brought from some unknown distance. During the middle ages Caen stone was frequently imported for building churches or other important architectural works. Before the Norman Conquest, however, most English buildings were of wood, so that, "to timber a minster," not to build a church, is the good early English expression of the chronicle. In chalk districts, at a later date, broken flints were often employed, and they give a mean appearance to the abbey ruins and churches at Reading, as well as to most of the older edifices at Brighton. Oxford, however, on the Oölite, is happily built of good native or imported stone. In modern times, London, standing in the midst of the brick-earth, has fallen a victim to the miseries of stucco,[3] until the Queen Anne revivalists have endeavored to restore an honest red brick; whereas Edinburgh, surrounded by excellent building-stone, has been able to do justice to its magnificent natural situation, and Aberdeen has clad itself in the stern but not unattractive gray and blue of its own solid granite. To the Caen stone, the Bath stone, and the Portland stone we owe half our cathedrals and abbeys, whose delicate tracery could never have been wrought in Rowley rag or Whin Sill basalt. The architecture of granite or hard limestone regions is often massive and imposing, but it always lacks the beauty of detailed sculpture or intricate handicraft. The marble lattice-work of the Táj or the "prentice's pillar" of Roslyn Chapel is only possible in a soft and pliable material.

Thus we see that agriculture and manufactures, art and science, are all largely influenced by geological conditions. Indeed, it would not be too much to assert that, after climate and geographical situation, geology is the greatest differentiating agent of national character. Every people is primarily what it is in virtue of the heredity it derives from the common ancestors of its whole stock; but, so far as it differs from other descendants of the same stock, the differences must mainly have been caused by those three great natural agencies, acting and reacting in conjunction with the original hereditary tendencies. The immense complexity of such actions and reactions renders them difficult to trace in detail; but the general principle which they illustrate can hardly be missed by those who read history with a wide and comprehensive glance.

1 - These little mollusks themselves abound upon chalky soils, and are found nowhere else, because they require large quantities of calcareous matter to form their banded shells, while other species with more horny coverings live on soils where less lime can be obtained. No snails can inhabit the limeless district of the Lizard in Cornwall. So minute are the interpendences between every portion of organic and inorganic nature.

2 - I owe this, with many other facts, to Mr. H. B. Woodward's interesting "Geology of England and Wales."

3 - Parker's cement, manufactured from the septaria of the London clay, is answerable for the outer coating of our West-end houses.

ÆSTHETIC FEELING IN BIRDS

There is no portion of Mr. Darwin's great superstructure which has been subjected to more searching criticism than his theory of sexual selection—the theory that beauty in animals is dependent, in part at least, upon the choice of brightly colored, ornamented, or musically endowed mates by one or other sex among all the more highly developed classes, such as insects, crustaceans, birds, and mammals. Not only have opponents argued strongly against the existence of such esthetic tastes in the lower animals as would account for the supposed preference for beautiful partners, but even many of those who accept the evolutionist hypothesis as a whole have declared themselves unable to give in their adhesion to this particular speculation. Professor Mivart has brought forward strong objections to the great naturalist's view, and Mr. A. R. Wallace has raised a counter-theory on the subject of coloration at least, which has done much to convince many wavering biologists, and to insure their rejection of the suggested cause as adequate for the production of that beauty which all alike recognize in the animal world. It seems to me, however, that a little too much stress has been laid upon the notion that comparatively advanced intelligence is necessary for the appreciation of beauty in the opposite sex. It is true that our own highly complex æsthetic feelings are largely composed of elevated intellectual and emotional elements; but it may perhaps be shown that aesthetic tastes quite sufficient for the production of the known results do actually exist in many cases, and consist almost entirely of very simple sensuous factors. In order for a butterfly or a humming-bird to admire its gorgeously appareled mate, it is not necessary that it should be capable of taking delight, like ourselves, in a Claude or a Rubens; it is enough that it should possess a nervous organization pleasurably affected by certain forms and colors in the same way as it is pleasurably affected by sweet fruits or the nectar of flowers. Nothing more than this need be postulated in order to establish the facts for which Mr. Darwin has contended with such wealth of illustration in the second part of the "Descent of Man."

It may be worth while, then, to examine a single large class of animals, in which the æsthetic nature is highly developed, for the purpose of discovering whether they do really afford proof of a sensibility to form, color, or musical sound. It must be remembered that even in our own race the sense of beauty in children, savages, and the uncultured classes, hardly rises above this simple level. We must not, of course, expect to find an appreciation of musical harmony, of imitative pictorial skill, of elaborate ornamentation, among birds or insects. We must be content if we see evidence of a love for red, blue, and yellow, for sweet perfumes and pleasant flavors, for symmetrical forms and simple patterns, for ringing notes and trilled resonances. The class of birds probably shows external marks of such tastes in a higher degree than any other; and, though many of them have been set forth by various writers elsewhere, it will perhaps repay the trouble to collect them into a single paper in order to show their bearing upon the general æsthetic sensibility of the class, as well as upon the specific question of sexual selection. For this purpose I shall first take for granted the fact of such selection, and afterward endeavor to justify it by analogy from known human practice.

Beginning with the lowest of the special senses, taste, we find ample evidence that very many birds have a strong liking for sugar. In confinement, canaries and parrots eagerly devour it in the manufactured form. In the wild state humming-birds, sun-birds, honey suckers, lories, and many other species, feed off the nectar of flowers, more or less mixed with insects. Mr. Webber, an American naturalist, found that the ruby-throats of the United States were attracted by a cup of sirup, and numerous other birds display a strong liking for the same mixture. Fruits, which have been developed especially to suit the tastes of birds, almost always contain an abundance of sugary juices; while the kernels within their stones are generally bitter, so as to prevent their winged allies from devouring the actual seed. Hence we may infer that all the vast tribes of toucans, hornbills, macaws, plantain-eaters, birds-of-paradise, and fruit-pigeons, possess a taste for sugar sufficiently strong to have produced the separate evolution of these sweet seed-coverings in a hundred different families of plants throughout the whole world. Indeed, the strength of the evidence thus afforded can not be overrated, when we remember that in every case the covering is a dead loss to the plant, except in so far as it aids the dispersion of seeds; and that it must have been developed over and over again in a thousand different cases by the action of the most widely different birds. It is impossible to believe that such a coincidence can be due to accident, impossible to doubt that it results from a genuine taste for sweet flavors.

There is even some reason to believe that birds care for and discriminate other tastes besides the fundamental distinctions of sweet and bitter. All the small birds in Jamaica are particularly fond of the little scarlet capsicums grown in gardens, and devour them so greedily, that the fruit has acquired the common name of bird-peppers. If we remember how very hard is often the almost horny covering of a bird's tongue, there is nothing remarkable in the fact that the pungency of the capsicum should be felt as an agreeable stimulant, probably having effects analogous to those of mustard, water-cress, or peppermint, with human beings. The oft-quoted liking of tropical pigeons for the nutmeg, with its aromatic coating of mace, points in the same direction. Parrots in captivity frequently display very decided preferences and antipathies in their food. Owls cannot be induced to taste meat in the slightest degree tainted. Again, all birds have a most accurate notion of the difference between ripe fruits and the unripe sour ones, besides carefully choosing the sunny side of peaches, pears, and apricots. The very frequency of distinct sapid principles in fruits would seem to favor the same supposition, as they have probably been acquired for the special allurement of particular species. Indeed, the more we consider the origin and nature of succulent fruits, the more does it become clear that they have been developed to suit the tastes of animals having essentially identical sensations with our own.

The case of the nutmeg leads us naturally on to the consideration of smell. Here we may conclude with great probability that the large class of aromatic fruits has acquired its perfume for the sake of attracting birds, especially when we recollect that flowers have acquired exactly similar perfumes for the sake of attracting insects. And although the possession of scent as a means of sexual allurement is rare among birds, being probably confined to the musk-duck and a few other species, yet it occurs frequently among butterflies, and is represented among mammals by the musk-deer, beaver, and many other ruminants or rodents. Curiously enough, the similarity of taste thus testified extends to the vegetal world in the case of the musk-plant; while even certain carnivores, such as the cat tribe, are extremely fond of "valerian, lemon-thyme, camomile, lavender, and many plants rich in essential oils." On the other hand, a good observer notes that cats have their dislikes, and he has often seen a tabby "smell at a fig-tree, and turn away with the disgusted air of a connoisseur." We have no such strong facts in the case of birds, but the frequency of perfumes in those fruits which depend upon them for the dispersion of their seeds, coupled with their total absence among most nuts, would lead to the conclusion that their likes and dislikes in the matter of smell are fully as marked.

But it is when we arrive at the sense of hearing that we come to the point where proper æsthetic feelings begin. It is quite impossible to doubt that birds are fond of musical sounds. The song of our own nightingales and linnets, the deep notes of the South American bellbird, the incessant cooing of the dove, the noisy chattering of the parrots, the ringing cry of the whippoorwill, all lead to the same conclusion. Here, again, these sounds are of precisely the same nature as those employed by the crickets, katydids, cicadas, and other musical insects, as well as by man himself in his vocal and instrumental music. Something of the same taste is displayed among the quadrumana by the howlers and other monkeys. But it is a noteworthy fact that a large majority of these presumably sexual calls, in birds, insects, and other animals, are true musical sounds, not mere noises. I have pointed out elsewhere the probable reason for this preference of pure tones in the case of mankind; and the same argument will apply, mutatis mutandis, to all other animals. But there is certainly a singular analogy in this respect between sounds and colors, most animals preferring the relatively pure and simple musical tones to confused noises; and the relatively pure and simple analytic colors, red, blue, green, and yellow, to confused mixtures such as brown, gray, and mud-color. At any rate, a bird evidently pays far more attention to the musical class of auditory perceptions than to mere noise. A canary will take no notice of ordinary confused sounds in a room; but, if one begins to chirp or whistle to it, it immediately responds with another chirp in emulation. So, too, when a piano or other musical instrument is played in the neighborhood of a singing bird, it will often show its recognition of the musical character by pouring out its very fullest flood of song, as if to conquer its unconscious rival. Of course, the singing-matches between birds themselves are too familiarly known to call for separate mention. It may be worth while, however, to notice that this love of musical sound exists even among certain reptiles; for I have often seen the common house-lizard of Jamaica listening with evident interest and attention to the playing of a piano, turning his head from side to side, and scampering away when disturbed, only to return again to the fascinating sound after a minute or two of hesitation.

The cases of the starling, the piping bullfinch, and the mockingbird, which can be taught to whistle a tune, show the same power still more highly developed. These instances prove not merely susceptibility to musical sounds, but also a capacity for distinguishing the harmonic intervals. It is stated that some birds, even in the wild state, display considerable knowledge of the musical scale; and a San Francisco naturalist is at present engaged upon a work in which he hopes to show that the human ear possesses in this respect merely a more highly developed form of the common vertebrate sensibility. When we reflect upon the purely physical and physiological basis which, as Helmholtz has taught us, underlies the musical intervals and the distinctions of harmony and discord, there is certainly no reason why they should not be perceived by all the higher animals alike, in a greater or less degree.

Considering, therefore, the evident susceptibility of birds to the simpler pleasures of music, and the interest which they show in it even apart from their domestic relations, there is no a priori difficulty in accepting the belief that their powers of song may have been developed by mutual selection, provided no adverse argument can be shown against the probability of such selection ever proving a cause of specific variation. To this last question, the question so ably raised by Mr. Wallace, I shall return on a later page.

Passing on to sight, we have first to observe the effects of mere light or brilliancy upon birds, apart from special effects of color or form. Now, birds certainly share with insects and many other creatures the common fascination for bright lights. "Owls and night-jars have been known to flutter against the window of a lighted room in the small hours."[1] In the tropics, where windows are more constantly left open, birds frequently fly into houses, attracted by a lamp or candle. The reflected light of a mirror is employed to draw down larks. Magpies delight in secreting diamonds, gold, silver, and other shiny objects. The bower-birds use shells, polished pebbles, and like brilliant odds and ends in the construction of their bowers. So, too, metallic iridescence occurs frequently in the feathers of beautiful species, notably in the humming-birds, sun-birds, peacocks, and other flower-feeding or fruit-eating classes. But even the far less brilliant crows, gulls, ducks, and doves show exquisitely burnished gloss or luster on their coats, often specialized upon particular portions of the plumage, and apparently betraying the action of sexual selection.

Of the love for color shown by birds, I have already treated so fully elsewhere, that it will suffice here briefly to recapitulate the main facts. The universality of bright hues in the fruits which depend upon birds for the dispersion of their seeds clearly shows that fruit eating species are attracted by red, blue, purple, and yellow; just as the analogous case of insect-fertilized flowers shows the preference of bees and butterflies for similar tints. Mr. Darwin has collected several instances of interest displayed by birds in colored objects-, and of the attractiveness which color evidently possesses in their eyes. Of these, the most remarkable cases are those of the bower-birds' and the hum^ ming-birds' nests. And the constant occurrence of very brilliant hues among flower-feeding species, such as humming-birds, sun-birds, lories, and barbets, or among fruit-eaters, such as toucans, fruit-pigeons, birds-of-paradise, and parrots, induces the belief that in these classes the exercise of the structures upon the search for food has led to the formation of a very strong taste for color, ultimately resulting in sexual modifications.

As for the harmony of color usually observable in birds, it must be remembered that our feeling of harmony probably depends upon the due intermission and alternation of sense-stimulants, and therefore ought naturally to be shared by us more or less definitely with all other animals having a like constitution of the eye. Now, the mammalian and avian eye being derived from a common ancestor, who already possessed a highly developed power of vision, we might reasonably expect that our feelings of harmony would be essentially identical; and this expectation is fully borne out both by the coloration of fruits and of birds themselves, which seldom or never present what we should regard as discordant coloring. Furthermore, as the most beautiful classes of birds are those which live perpetually among tropical flowers and fruits, in the most beautiful forests or meadows, surrounded by exquisite insects and reptiles, and forever exercising their vision upon the most diversely colored environment in the whole world, it would seem far from impossible that their chromatic sensibility is even more highly developed than that of average humanity, and therefore that harmony or discord of colour would bear a relatively greater importance in their eyes than in those of any human being except the most artistically endowed. This conclusion will doubtless sound strange and even grotesque to those who are always accustomed to postulate for man a kind of absolute supremacy in the scheme of Nature; but it appears to me almost as obvious and as simply accounted for as the superiority of scent in the dog and the deer, or of distant vision in the eagle and the vulture. Lastly, it may be noted that much of the beauty of birds, as of insects, fruits, and flowers, is due to the delicate gradation of tints which they display. But in all natural products such gradation is an almost necessary result of the mode by which they have been evolved. It is only in human manufactures, where pigment is laid on with a brush or stamp, that colors can be placed in crude juxtaposition to one another, giving rise to the worst form of chromatic discord. Doubtless our native feeling of dislike to such discords, based upon their immediately fatiguing effect upon the nerves employed, has been heightened intellectually by the knowledge that they differ so widely from the dainty gradations to be found in the handiwork of Nature. Besides being sensuously recognized as discordant, they are intellectually recognized as inartistic. Thus a large part of our art-progress has consisted in an advance from the harsh and monotonous fields of primary red and blue, divided by very hard and definite lines, which we find in Egyptian painting, to the faithful representation of graduated tints and shades which appears upon a modern canvas. But in the petal of a rose, the ray of a daisy, the wing of a butterfly, the tail-covert of a peacock, such gradual merging of tint in tint could hardly fail to occur spontaneously, as a product of evolution; while the comparatively definite marking off of special spots or lines, as in some orchids and other flowers, could only present itself as a result of very intense competition between species, carried on under highly complex conditions. The views set forth by Mr. Bates upon the progressive modification of patches or regions in a butterfly's wing, and by Mr. Darwin and Mr. Wallace on the feathers of the peacock and the Argus-pheasant, though widely differing as to the particular mode of their evolution, yet alike convince us that the inevitable result must be just such a graceful running together of contiguous colors as we actually find to obtain in every case.

Lastly, we arrive at the sensibility to form, symmetry, arrangement of patterns, and the like higher sensuous aesthetic feelings, which remains in the eyes of many the chief stumbling-block in the way of accepting the theory of sexual selection. The pleasure derived from sweet tastes and fragrant perfumes is so purely sensuous that nobody doubts its universal existence among all the higher animals. The pleasure derived from musical sounds and bright colors, though more intimately bound up in the human consciousness with intellectual and higher emotional elements, yet contains so large a factor of mere sensuous stimulation that we can easily conceive of it as appealing to the ears and eyes of insects and vertebrates. But the still higher pleasure derived from graceful curves, symmetrical ornamentation, and elaborate tracery is so largely made up of intellectual feelings, and so largely supplemented in our own case by associations of costliness, human handicraft, or imitative skill, that we find it hard at first sight to believe in the existence of similar feelings among pheasants of the Indian jungle, antelopes of the African plains, or monkeys of the Brazilian forest. Even here, however, a little consideration may convince us that the æsthetic appreciation of form and its connected varieties is not necessarily above the narrow intellectual faculties of the higher vertebrates and articulates at least.

In the first place, if we look at the human race itself we shall find that a comparatively high susceptibility to form occurs even among very low races. Indeed, most exquisite patterns are produced by savages whose taste in color is apparently far less developed than that of parrots, humming-birds, and fruit-pigeons. The tattooed tracery of the Polynesians and many other savage tribes presents beautiful designs of which even a European decorative artist need not be ashamed.

The New Zealand canoes, the paddles and clubs of the Admiralty-Islanders, the shields of the Zooloos, are all most graceful in their shapes and most daintily wrought with interlacing patterns in carved work. Calabashes, cocoanuts, ostrich-eggs, and other early vessels are always cut in sections which exactly coincide with the demands of the most developed taste. The huts of savages are generally square, circular, or oval in shape, neatly wattled at symmetrical distances. The earliest architecture consists of regular stone rings, avenues, tumuli, and other definitely shaped monuments. Dr. Schweinfurth's "Heart of Africa" contains pictures of pottery as beautiful as anything ever produced in Greece or Etruria, stools, chairs, and other furniture as gracefully shaped as anything ever wrought by a Renaissance carver, and villages as prettily arranged after their simple fashion as the architects of the Parthenon or Cologne could have arranged them. If we look back in time, we find the stone hatchets and arrow-heads, not only of the neolithic but even of the palaeolithic age, carefully symmetrical in shape, and that at a time when the extra labor of chipping the flints into comeliness must have entailed a considerable waste of human or half-human energy. At the same early date we find fossil shells, symmetrical bones, teeth, and other like objects, already drilled to serve as necklaces or other ornaments, which analogy with the similar ornaments now in use would lead us to believe were symmetrically strung together into definite patterns. Indeed, the more we look at the products of the very lowest savages and the very earliest men, the more shall we be convinced that they possessed in the germ all those aesthetic feelings which have finally developed our existing architecture and other decorative or semi-decorative arts.

Again, we cannot fail to be struck by the fact that man has always employed for ornamental purposes exactly those very appendages of animals which, if the theory of sexual selection be correct, have been produced by the animals themselves as ornamental adjuncts. The feathers of peacocks, the plumes of the ostrich and the bird-of-paradise, the antlers of deers, the horns of antelopes, the tusks of elephants, mammoths, and musk-deer, the striped, spotted, or dappled skins of mammals, all these have been used from the earliest periods as materials for decoration by mankind. Exactly the same curls, twists, and patterns which seem to please the eyes of animals are known to please the eyes of man, even in his lowest developments. If these ornaments were not produced because the creatures themselves found them beautiful, at least they are the same as those which would have been produced had the taste of such creatures coincided in the main with that which runs throughout the whole of humanity, from the most degraded savage to the highest artist.

Moreover, part at least of the pleasure of form probably has a purely sensuous origin. The superiority of curved lines to straight, of the waving or sinuous contour to the angular, is apparently connected with the muscular process in the act of vision. Hence there is no reason why it might not be felt by intelligent animals, just as we know that it is felt, and acutely felt, by hardly more intelligent men.

Similar conclusions are forced upon us if we look at the nature of the supposed ornaments themselves. They are almost always, like the horns of several ruminants, the tail-coverts of the peacock, and the lappets or crests of many birds, apparently devoid of any functional use whatsoever, unless that use be the attraction of the opposite sex. They are also marked by the extreme definiteness of their shape, color, or sculpture—a definiteness which never occurs in similar structures among the lower animals. For though some echinodermata, as for example the sea-urchins, are very beautifully and regularly marked, yet their markings are purely dependent upon the structural arrangements of the animal, and cannot generally be detected till after death. So, too, the shells of many mollusca, such as scalaria and the murices, are very beautifully sculptured; but this sculpture is structurally necessary for the animal, and apparently depends entirely upon the shape and markings of the mantle. Among birds, however, as among the ruminants, all the structures ascribed by Mr. Darwin to sexual selection are marked by a kind of definiteness, quite unconnected with ordinary functions, which it is difficult to describe in words, but which can immediately be felt if we compare the coloration of a peacock with that of a sea-anemone or a medusa. The former is perfectly definite without being obviously connected with structure; the latter is very indefinite, and yet bears a clear relation to the general shape of the animal. This combination of great specific distinctness with little apparent functional value appears to me the genuine hall-mark of organs due to sexual selection.

There is even some little external evidence in favor of a love for symmetry among birds. The nests of weaver-birds and many other species, as well as the bowers of the bower-birds, display a considerable taste for orderly arrangement. For one must remember that the building of such nests, though doubtless instinctive and inherited, is not a mere organic process, like the secretion of a molluscan shell; it is as much an art as the building of a honeycomb or of a savage hut. The flight of birds in play, the antics of many humming-birds, the strange eddyings and aërial evolutions of several other species, all approach very nearly to our own idea of dancing. I am almost afraid to hazard the observation, yet on the other hand I cannot avoid risking it, that the attitudes taken up by the turkey-buzzards or John crows of the West Indies upon the tops of houses frequently seemed to me intentionally symmetrical. I have observed them sitting in every variety of position—one at each end of a long roof; one at each of the two points half-way between ends and middle; three arranged in either of these forms, with one in the middle; five arranged in the order C, B, A, B, C, etc. If any other observer can supplement this experience, which I record with great diffidence, I shall be very glad.

Taking for granted, then, this appreciation of form and symmetry, we shall find that it has produced many notable effects in the world of birds. To it, apparently, we owe the crests of cockatoos, pigeons, herons, and a hundred other species; the wattles, combs, hackles, and lappets of the gallinaceous birds; the beaks of toucans, hornbills, and cassowaries; the wonderful marking of the peacock and the Argus pheasant. Anyone who wishes really to understand the immense variety of ornamentation which has thus resulted should pay a visit to the ornithological rooms in the British Museum, and observe the innumerable devices for attracting attention which exist in almost every order of birds. Perhaps the familiar lyre-bird offers the very finest example of all, so far as beauty of form and symmetry of arrangement are concerned. It is specially noticeable, however, that in almost every case the decorations are lavished on the very same parts on which they would have been bestowed by human taste.

If, then, we put together all the scattered indications thus afforded us, if we consider the taste for sweet food and delicate perfumes, the song of the nightingale and the graceful movements of the swan, the metallic colors of the flower-feeders, the exquisite hues of the fruit eaters, the varied plumage of the birds-of-paradise, the beautiful nests and bowers, the habit of abstracting brilliant objects, the universal loveliness of shape or tint throughout the whole class—we can hardly doubt that birds, as a whole, possess aesthetic endowments of a very high order. Let us proceed to consider the general bearings of these views upon the question of sexual selection.

Mr. Herbert Spencer, in a very remarkable essay upon personal beauty, has shown that in the human race we regard as beautiful, on the whole, just those personal peculiarities which are, roughly speaking, the external marks of fitness for the conditions of human life. More especially do we admire those points which bespeak a physique adapted for the duties of paternity and maternity. We dislike excessive leanness or excessive fat; a sallow or a bloated complexion; deformity or extreme departure from the normal type. On the other hand, we like in man robust and muscular limbs, an erect carriage, an open chest, a virile development of beard and whiskers, with all the other outward signs of health and strength. We like in woman a womanly and tender face, a fine and well-developed figure, and all the other outward signs of health, and more especially of healthy maternal capacities. We like in both sexes an abundant crop of hair, clear and bright eyes, white and well-set teeth, red lips, and cheeks which show a good and sound circulation; we like an expression which betokens good humor, moral qualities, and refinement; lastly, we like a face which indicates intellectual power and ability to succeed in the highly complex struggle for life in the midst of which our lot is cast. One or other of these points we may occasionally waive in consideration of other special claims; but if anybody asks in the abstract whether we prefer a stunted physique to well-grown limbs and muscles; a flat-chested woman to one with a finely-proportioned bust—unhealthy and sallow skin to a clear complexion; a sour-looking, mean, or brutal face to a bright, joyous, open, and honest countenance; silly or idiotic features to an expression full of liveliness and intelligence—there can be but one answer possible. Leaving out of consideration for the present all other elements of the involved and complex problem, we may conclude that beauty, from one point of view at least, consists for each species in the outward signs of specific adaptation to specific necessities.

On the other hand, beauty also consists from a different point of view of stimulation by a certain relatively fixed number of external stimulants—musical sound, brilliant light, analytic colors, curved shapes, symmetrical arrangements of form, etc.—which appear to act directly upon the nervous system. This is clearly the view which Mr. Darwin implicitly accepts, especially with regard to tone and color. The facts at which we have briefly glanced above respecting the aesthetic feelings in birds, and the beauty of the birds themselves, take for granted some such theory of the aesthetic faculty. How are we to find a reconciliation between this view and that of Mr. Herbert Spencer?

I believe the true clew has been given us by Mr. A. R. Wallace, in the able essays on "Color in Plants and Animals" which originally appeared in "Macmillan's Magazine," and were afterward reprinted in his work on "Tropical Nature." It is true that Mr. Wallace utterly rejects sexual selection as a vera causa, and substitutes for it several separate minor modifications of natural selection; yet it seems to me that a compromise between his view and the two other views of Mr. Darwin and Mr. Herbert Spencer would more really represent the actual state of the case in nature. Or, to put it more correctly, the three ideas are not in reality contradictory or even opposite, but are rather different and complementary aspects of one and the same fundamental truth.

Beauty in the abstract and for all species, as it seems to me, consists of pleasurable stimulation of the higher sense-organs. Such pleasurable stimulation must, on the average of cases, be given rather by brilliancy than by dullness; rather by analytic colors than by confused hues; rather by curved or flowing forms than by angularity; rather by musical sounds than by mere noises. But beauty relatively to the particular species, and especially as regards the sexual relation, must be largely due to special inherited tastes, doubtless ingrained and physically registered in the nervous system, leading the animal to derive pleasure from the typically healthy and normal form of the opposite sex. For, if any individual possesses divergent tastes, they must either be for relatively unhealthy and typically defective forms, in which case they will tend to be promptly suppressed by natural selection; or for neutral or improved forms, in which case they will help to give rise to new varieties, ultimately culminating in separate species. Such divergent tastes seem to be shown in all large dominant families, such as the humming-birds, where specific variation and ornamentation have been carried out to a very great extent. But all such divergent fancies must themselves tend to become distinctly fixed for purposes of specific identification; and we find as a matter of fact that each species does readily recognize its mates, even when the differences between closely allied species are only very slight.

Now, this special hereditary liking for a particular form and type will not interfere with the general love for color, brilliancy, sweet tones, and perfumes. Accordingly, wherever the circumstances which give rise to a taste for these sense-stimulants exist, it would naturally follow that the taste would help to determine the choice of mates. But, again, as Mr. Wallace has fully shown, the most vigorous individuals would usually possess the most highly developed ornaments, the brightest colors, the largest scent-glands, and the loudest or most musical voices. Hence the very animals most likely to be sexually selected are also, on the average, those most likely to be naturally selected. Yet sexual selection really differs from natural selection, in that it gives a special direction to the ornamentation. For example, one can hardly believe that mere masculine vigor will account for the gorgeous and positively inconvenient plumage of the bird-of-paradise, nor for the exquisite coloring of the peacock, nor for the extremely ungainly air-bladders of many insects. It is quite easy to conceive that the general vigor implied by the possession of these extended ornamental adjuncts may have helped their possessors in the general struggle for life; but it is hardly possible to believe that they could have reached their present definite development without the aid of sexual selection. In short, where an ornament, or what seemed to any particular individual an ornament, proved hurtful to the race, it would be eliminated by natural selection; but where it proved neutral it would be spared, and if it coincided with advantageous qualities it would be further developed. Yet, even if only neutral, sexual selection alone would give it an extra chance, and, as it would doubtless be correlated on the one hand with certain special tastes and habits, and on the other hand with certain slight modifications of structure, it would doubtless succeed on an average of cases in producing a new species.

The familiar facts of human beauty will probably serve to make this reconciliation of the conflicting views a little clearer. Man of course admires in the abstract bright colors, brilliancy, musical notes, graceful curves, and symmetrical form. But, as applied to the human face and figure, he admires these in certain special and typical arrangements. Thus, while our general love for color leads us to prize golden hair, we do not like a sallow complexion; while it leads us to see beauty in rosy cheeks and red lips, which are signs of a healthy circulation, we do not admire the same redness in the nose, where it is usually a result of dyspepsia or dissipated habits, either of which is bad for the race at large. Again, though we admire pearly teeth, clear eyes, and a white skin, all of which are obviously the external marks of useful properties, we do not admire white cheeks, which are the external mark of weakness or anæmia. Similarly, our idea of beauty demands that the figure should neither be too fat nor too thin, but should possess that graceful development of all the muscles which is the outward symbol of ability to move and act with ease and effect. If any large number of persons were ever actuated by opposite tastes, if they preferred pale cheeks and lips to rosy ones, thin and haggard faces to full and rounded ones, weak and angular limbs to strong and graceful ones, a flat and undeveloped chest to a fine and healthy bust, then they and their taste must rapidly die out through the inferior physique they would hand on to their descendants. And as every individual is himself the product of countless thousands of prior individuals, all of whom have been in the main successful in the struggle for life and the search for mates, it must follow that he will have inherited from them, on the average, a healthy taste for that particular arrangement of limbs and features which best suits the essential conditions of the species. Not, of course, that he will consciously recognize this fact in most cases; but the mere presentation of such a typical combination will instinctively rouse in him, through the organized correlation of nervous centers, the hereditary feeling of beauty. Hence this feeling will probably be most strongly aroused in each species by the sight of the sex which in that species has undergone the greatest differentiation through sexual selection: just as we know that the feeling is most strongly aroused in mankind by the beauty of woman. On the other hand, we are still able to perceive, when we look at a peacock or a humming-bird, that, thought his specific hereditary feeling is absent, yet the strength of the purely abstract elements—color, brilliancy, symmetry, form, and minute workmanship—is so unusually great that we have no hesitation in pronouncing them also beautiful after their kind.

If, then, we admit the reality and potency of sexual selection, in however modified a form, it must follow that birds, being on the whole the most ornamental of all classes in the animal world, are also the most æsthetic, with the exception of man. It might, at first sight, seem that consistency would demand the sacrifice even of this exception; but a moment's reflection will disclose an important difference between the two cases. Man possesses the active power of direct artistic creation; the birds only possess the passive power of selection from among the forms produced for them by Nature. The ordinary workman who selects his wife partly or wholly on the ground of beauty, thereby does something toward perpetuating and improving the beauty of the race; he stamps the impress of his taste upon future generations; but such mere passive choice differs widely from the ability to depict or create on canvas such a beautiful woman. In this way, the actual loveliness of birds may lead us somewhat to over-estimate their aesthetic sensibility; for, though within their own species they may be capable of distinguishing between comparatively minute shades and degrees of beauty, just as we can distinguish between such minute points in human faces as would doubtless absolutely escape the notice of any other animal, it is yet improbable that they would be equally discriminative outside the limits of their own species. Again the principle of "gradation of characters" necessitates certain artistic effects in their plumage which they themselves may be only half able to admire. So, too, the necessarily symmetrical arrangement of the two sides of the body and the mode of growth of feathers may often have helped, unintentionally, as it were, in producing the total effect. In other words, it may well be that the birds, while selecting their partners on the ground of bright color, exceptionally long plumes, and other ornamental characters which they could understand and admire, may have succeeded in producing harmonies of tone, delicate gradations of tint, and other similar effects which they could not understand or admire, or at least could only admire very partially.

Yet, after making all allowances for possible reading in of human feelings, it may probably be asserted with safety that the actual appearance of birds entitles them to rank, on the whole, higher in the aesthetic scale than any other animals except man. Whether we look at their graceful shapes, in the swan and the heron; their beautiful plumes, in the ostrich and the bird-of-paradise; their exquisite color, in the sunbird and the lory; their ornamental crests and lappets, in the humming-birds, the pigeons, and the parrots; or their song in the linnet, the mocking-bird, and the nightingale—we must confess that they give extraordinary evidence of a taste for all that man considers lovely or artistic. And this is just what we might expect from their free mode of life, their rapid motion, their highly developed senses, their comparative freedom from enemies, their long and almost uninterrupted rivalry between themselves for the possession of their mates. Especially should we expect this splendid outburst of aesthetic sensibility exactly where we find it in its greatest glory, among the flower haunting and fruit-eating species of the Brazilian forests, the Indian jungles, and the Malay Archipelago. Surrounded for generations and generations by gorgeous orchids and trumpet-creepers, from which they sucked the stored-up nectar, by gleaming purple or golden fruits, by burnished beetles, metallic butterflies, bronze-scaled lizards, and coral snakes, their prey or their enemies, exercising their eyes perpetually in the search for food among the exquisite objects of their environment, and safe from almost all foes except those of their own class, tropical birds have naturally developed the most gorgeous and the most perfect forms and colors in the whole animal creation. And, above all, they have stamped the mark of their peculiarly high aesthetic feelings upon their own shapes by the wonderful definiteness of their patterns and their ornamental adjuncts, nowhere equaled, save in the most perfect decorative handicraft of man himself.

1 - This, with several other instances, I take from an interesting article on "The Senses of the Lower Animals," in the "Quarterly Journal of Science" for July, 1878.

All the higher processes of evolution are necessarily so complex in character that we can really deal with only a single aspect at a time. Hence, in spite of the rather general title which this paper bears, it proposes to treat of æsthetic evolution in man under one such aspect only—that of its gradual decentralization, its increase in disinterestedness from the simple and narrow feelings of the savage or the child to the full and expansive æsthetic catholicity of the cultivated adult. We have to trace the progress of the sense of beauty from its first starting-point in the primitive sensibilities of the race or the individual to its highest development in the most refined and advanced of European artists.

To do so, we must first find this starting-point itself. What is the center from which the widening circle of æsthetic sensibility takes its departure? In other words, what is the primitive source of the appreciation of beauty? Putting the question into a concrete form, what objects did man, as a whole, and does each man in particular, first find beautiful? If we look at a cultivated European, we see that he derives great æsthetic enjoyment from contemplating the sunset clouds, the green trees, the lakes, rivers, and waterfalls, the flowers, birds, and insects around him. But, if we look at a savage or a child, we see that for the most part they care for none of these things. We might almost conclude, on a hurried glance, that they had no sense of beauty whatsoever. Yet, when we examine them a little more closely, we find that there are many objects to which they do apply some such word as pretty, the symbol of the simplest æsthetic appreciation. If we can discover the limitations of these earliest æsthetic objects, we shall have solved one of the most important fundamental problems in the theory of beauty.

The settlement of such fundamental problems seems to me an indispensable preliminary to the construction of a scientific doctrine of æsthetics. When professors of fine art discuss the principles of beauty, they are too fond of confining themselves to the very highest feelings of the most cultivated classes in the most civilized nations. The mere childish love of colors, the mere savage taste for bone necklets and carved calabashes, seem beneath their exalted notice. Nay, more, we constantly find them accusing one another of having no feeling for beauty, or at least very little. Thus we see Mr. Ruskin and Mr. Poynter each mutually denying the other's powers of appreciation. But the psychological æsthetician cannot confine his attention to such exceptional and highest developments of the love for beauty as engage the whole interest of these artistic critics. He must look rather to those simpler and more universal feelings which are common to all the race, and which form the groundwork for every higher mode of aesthetic sensibility. It is enough for him that all village children call a daisy or a primrose pretty: he need not go far afield to discuss the peculiar specific merits of a Botticelli or a Pinturiccio. Hundreds of thousands, who would stare in blank unconcern at a torso from the chisel of Phidias, can love and admire "the meanest flower that blows," with something not wholly unlike the welling emotions of a Wordsworth. Indeed, one is often inclined to fancy that the truest lovers of beauty in nature, or in the works of man, are not always those who can talk most glibly the technical dialect of art-criticism.

If we wish to hit upon the primitive germ of æsthetic sensibility in man, we cannot begin better than by looking at its foreshadowing in the lower animals. There are two modes of aesthetic feeling which seem to exist among vertebrates and insects at least: the first is the sense of visual beauty in form, color, or brilliancy; the second is the sense of auditory beauty in musical or rhythmical sound. The former of the two modes I have endeavored in part to illustrate in my little work "The Color-Sense": the latter has been admirably treated by Mr. Sully in his valuable essay on "Animal Music," which appeared in the "Cornhill Magazine" for November, 1879. Now, if we look at the manner in which insects, birds, and mammals apparently manifest these presumed æsthetic feelings, we shall see that they are very restricted and limited in range. Animals never seem to admire scenery, or foliage, or beautiful creatures of other species. They do not appear for the most part to care greatly for human music, or for any sounds other than those uttered by their own kind. They do not even show any marked aesthetic enjoyment of the lovely flowers and fruits whose tints, as Mr. Darwin teaches us, are mainly due to their own selective action. But, if our great biologist is correct in his reasonings,[1] they do very distinctly display their admiration for the beautiful forms, colors, and songs of their own highly decorated or musical mates. The facts on which Mr. Darwin bases his theory of sexual selection thus become of the first importance for the aesthetic philosopher, because they are really the only solid evidence for the existence of a love for beauty in the infra-human world. Granting the truth of his views (on which I for one have no shadow of doubt now remaining), we have good proof of a taste for symmetry and curved form in the magnificent tail of the lyre-bird, in the wedding plumage of the whydah-bird, in the twisted horns of the kudu antelope; of a taste for color and luster in the gorgeous train of the peacock, in the metallic necklets of the humming-bird, in the exquisite wings of tropical butterflies, in the bronze and gilded armor of the rose-chafers; lastly, of a taste for musical sound in the stridulation of the cicada and the house-cricket, in the deep notes of the bell-bird and the howler monkey, in the outpoured song of the linnet, the sky-lark, and the nightingale.

This close restriction of the æsthetic feeling to those objects which most nearly concern the individual, and through him the species, is only what we should naturally expect among the lower animals. We could hardly fancy them interesting themselves in anything so remote from their own personal wants as the rainbow or the sunset, the blue hills and the belted sea. They and their ancestors before them could not have gained any advantage by turning aside their attention from the practical pursuit of food or mates, to the otiose contemplation of that which profiteth nothing. Our own disinterested love for things so distant from our substantial needs has arisen gradually through a long process of ever-widening sympathies and ever-multiplying associations. But two things the insect, the bird, or the mammal could notice, and gain an advantage for itself or its race by noticing. It could pick out by its eye the forms and colors of edible foodstuffs among the unedible and relatively useless mass of foliage upon—earth the red berry or blossom from the green leaves, the fat white grub from the brown soil, the lurking caterpillar from the stalk whose lines and hues it so exactly imitates. It could distinguish by its ear the chirp of the savory grasshopper from the click of the hard or bitter beetle, the pretty note of the harmless sparrow from the deep cry of the dangerous hawk or the greedy jay. Thus eye and ear alike became educated among the superior articulates and vertebrates, in anticipation, as it were, of their higher aesthetic functions.

In the choice of mates, however, the powers so gained were exercised in a way which we cannot consider as falling short of the true aesthetic level. Even the lowest animals (among those in which the sexes are different) seem instinctively to distinguish their fellows from all other species. In the higher classes, where the eye and ear have been so educated as to discriminate minutely between various forms, colors, shades, and notes, the instinct must almost certainly operate through the senses of sight and hearing. Even among those races of insects, birds, and mammals in which no distinct marks of sexual selection exist, I believe the sight of beautiful members of their own kind must necessarily excite pleasurable feelings worthy of being ranked in the aesthetic class. In other words, I believe every crow must think its own mate beautiful—not merely inferentially pleasant, but in the truest sense beautiful. There must be, it seems to me, such an intimate correspondence between the needs and the tastes of each species, that the sight and voice of a healthy, normal, well-formed mate must have become intrinsically pleasing for its own sake, as well as indirectly for its associations. The nervous centers of each species must, I conceive, be so constructed hereditarily as to answer congenitally to certain typical shapes and sounds often experienced ancestrally, and always with ultimate benefit to the race. Though the emotions require experience of the object to arouse them, when the object occurs the emotions naturally arise. Just as man has special cerebral structures—existing, though dormant, even in deaf-mutes—for the perception and production of human language, so, I cannot but believe, every species of higher animal has special cerebral structures, with special corresponding blank forms of perception, for the intellectual recognition and appropriate emotional reception of its fellows and its mates. These feelings are innate in the sense that they occur spontaneously at sight of the proper objects. When Miranda falls in love at first sight with Ferdinand, the only young man she has ever seen, it seems to me that the poet has truly depicted a genuine psychological fact. At any rate, it is indubitable that, so far as man is concerned, the human voice has certain points of emotional and technical superiority over every other kind of musical instrument, and that the beauty of woman and of the human form is now and must always remain the central standard of beauty for all humanity.

The heart and core of such a fixed hereditary taste for each species must consist in the appreciation of the pure and healthy typical specific form. The ugly for every, kind, in its own eyes, must always be (in the main) the deformed, the aberrant, the weakly, the unnatural, the impotent. The beautiful for every kind must similarly be (in the main) the healthy, the normal, the strong, the perfect, and the parentally sound.[2] Were it ever otherwise—did any race or kind ever habitually prefer the morbid to the sound, that race or kind must be on the high-road to extinction. The more every individual shares the healthiest tastes of its kind, and puts them in practice in the choice of a mate, the more is he or she insuring for descendants a healthy and a successful life whereby it hands on its own sound taste to future generations. But, besides this fundamental typical beauty—the beauty which consists in full realization of the normal specific form—there is another source of personal beauty on which sexual selection may act, and through which it has produced the greater number of its most striking effects. This source may be found in the exercise of tastes otherwise acquired upon relatively unimportant details of form, color, or musical abilities. The taste for bright hues, acquired through the search for food in blossoms, berries, or brilliant insects, may be transferred to the search for mates, so that those mates will be most preferred which happen to vary most from the original typical coloration in the direction of more brilliant hues. The taste for musical sound, implied, as I have elsewhere tried to show on the lines laid down by Helmholtz, in the very structure of the auditory apparatus (at least in birds and mammals), may be exercised in the preference given among birds to the sweetest or the loudest singers. Unimportant ornamental points may thus be constantly developed by continual selection of small gradations, when they do not interfere with the general efficiency of the organism, till at length we get such highly evolved aesthetic products as the waving plumage of the bird-of-paradise, the sculptured antlers of the gazelle, and the varied song of the mockingbird. And since, as Mr. Wallace has shown (he himself believes in opposition to, but I rather fancy in confirmation of, Mr. Darwin's theory), these ornamental adjuncts or faculties are most likely to coexist with the highest sexual efficiency, it must happen that in the main sexual selection and natural selection will reënforce one another, the strongest and best being always on an average the most beautiful, and hence the most pleasing to all possible mates.

In this way, I take it, a sense of beauty in the contemplation of their own mates must have grown up among all the higher animals, and must have became strongest and most discriminative among those whose mates have undergone the greatest amount of ornamental differentiation. And as the secondary differences between man and woman as to beard, hair, and features, are greater than between the two sexes of almost any other quadrumanous animal, we may conclude that man's aesthetic appreciation of beauty in his own species has always been very considerable. Of this æsthetic appreciation, the secondary differences in question are at once the proof, the cause, and the effect. For, in the constant action and reaction of heredity and adaptation, it must happen that the greater the original taste, the more will it be exerted in the choice of mates; and, the more it is exerted in each generation, the greater will be its effects, and the more will the taste be strengthened in all future generations.

This, then, would seem to be the primitive starting-point of which we are in search. Man in his earliest human condition, as he first evolved from the undifferentiated anthropoidal stage, must have possessed certain vague elements of æsthetic feeling: but they can have been exerted or risen into conscious prominence only, it would seem, in the relation of primeval courtship and wedlock. He must have been already endowed with a sense of beauty in form and symmetry, a sense which, in spite of its wide expansion and generalization in subsequent ages, still attaches itself above every other object, even with Hellenic or modern sculptors, to the human face and figure. He must also have been sensible to the beauty of color and luster, rendered faintly conscious in the case of flowers, fruits, and feathers, but probably attaining its fullest measure only in the eyes, hair, teeth, lips, and glossy black complexion of his early mates. And he must have been moved, as Mr. Darwin argues, by musical tones and combinations, though chiefly in the form of human song or rhythm alone. In short, the primitive human conception of beauty must, I believe, have been purely anthropinistic—must have gathered mainly around the personality of man or woman; and all its subsequent history must be that of an apanthropinization (I apologize for the ugly but convenient word), a gradual regression or concentric widening of aesthetic feeling around this fixed point which remains to the very last its natural center. By the common consent of poets, painters, sculptors, and the World at large, the standard of beauty for mankind is still to be found in the features and figure of a lovely woman.

Probably primitive man admired his pre-glacial Phyllis or Neæra, admired himself, and perhaps also admired his fellow-man. So far as I can learn, there are no savages so low that they do not discriminate between pretty squaws or gins and plain ones, between handsome men and ugly ones. Our own children appear to me to make the distinction among their playmates from a very early age. And, in both cases, I am satisfied that their judgment in the main agrees with our own.[3] But it does not seem likely that primitive man took much notice of scenery, of organic beauty as a whole, or even very largely of beauty in flowers, berries, butterflies, and shells. Yet there was an obvious link, a simple stepping-stone, by which nascent aesthetic feeling might easily pass from the one stage to the other. That link is given us in the love for personal decoration.

Not only does every unsophisticated man wish to find a pretty mate, but he also wishes to look to advantage in her eyes and those of his rivals. Similarly, every woman wishes to look pleasing toward all men. The most naked savages take immense pains with their fantastic coiffures. Even birds display their beauty to the best advantage, and sing in emulation with one another till their strength fails them. But birds and mammals generally go no further than this: man can take one step in advance, and add to his natural beauty, or conceal his natural defects, by borrowed plumes. So the earliest evidence of derivative æsthetic feeling which we possess is that of the personal ornaments worn by palaeolithic men. Perforated shells, apparently used for necklaces; teeth of deer and other animals; pebbles of rose-quartz and other ornamental stones; wrought pieces of bone or mammoth ivory—all of them obviously intended for personal decoration—are found in the earliest cave-dwellings and rock-shelters. Feathers and flowers we cannot of course expect to find in such situations; but we can hardly doubt, from the analogy of almost all modern savages, that palaeolithic men must have used them as much as they used those other decorative objects. Now, the fact that any such shells or plumes are sought as ornaments proves of course that they were first admired; but the vague admiration originally bestowed upon them would naturally be much quickened and increased by their employment for the decoration of the person. From being vague and indefinite it would become vivid and purposive. Our own children and modern savages take comparatively little interest in flowers in the abstract, flowers as they grow upon the bush or in the field: but they begin to admire them when they pick them by handfuls, and still more when they are woven into a wreath, arranged in a bouquet, or stuck into the hair. Nay, is not this ultimate decorative intent one of the chief raisons d'être for many of our European conservatories and florists' shops? Is not a camellia largely admired because it looks so well in a ball-dress, and a stephanotis because it fits so easily in a button-hole? And is it not a fact that many of our ladies and most of our seyants admire artificial flowers, with all their stiffness and vulgarity, far more genuinely than they admire living roses or lilies-of-the-valley? We have all known women whose most real æsthetic feelings were obviously aroused by a bonnet or a head-dress.

Flowers are very favorite decorations with the South-Sea Islanders, and those who have read Miss Bird's and Mrs. Brassey's pleasant accounts of their stay among the Polynesians must have noticed the air of refinement, the vague aesthetic atmosphere thrown over the whole story by their profuse employment of tropical blossoms upon all occasions. Feathers, symmetrically arranged, were the ordinary head-dress of the North American Indians; and they were woven into splendid cloaks by the Hawaiians. Corals, pebbles, precious stones, gold and silver jewelry, cowries, wampum beads, furs, silks, and so forth, follow in due order. Ochre and woad, for dyeing or staining the body, are employed from a very early period. Henna, indigo, and other cosmetics come a little later Among many existing lowest races, the only sign of aesthetic feeling, beyond the sense of personal beauty and the very rudest songs or dances, is shown in the employment of dyes or ornaments for the person. Such are many of the Indian Hill tribes, the Andamanese, the Digger Indians of California, and the Botocudos of Brazil. The Bushmen, and to a less extent the Australians, generally ranked in the lowest order, reach a decidedly higher æsthetic level.

In most savage communities, the men, not the women, monopolize the handsomest costumes, which are worn as marks of distinction, not merely as ornaments. But the former use must be necessarily derivative and secondary, not original. Mr. Herbert Spencer has gathered together a large and interesting collection of cases in his "Ceremonial Institutions" (chapter ix). Nevertheless, the original aesthetic intent of most of such decorations is obvious from the fact that they are universal among women, whenever they do not arise from the habit of trophy taking, as with the use of flowers with the Polynesians generally. So, too, tattooing and other mutilative practices, originally subordinative in their intention, becoming at last merely æsthetic, are prized by women as increasing their natural attractions. Everyone must remember the plea of the New Zealand girls, quoted by Mr. Darwin, who answered the remonstrances of the missionary against tattooing by saying, "We must have just a few lines upon our lips, or else when we grow old we shall be so very ugly." Similarly, Central African women admire their own pelelé, the piece of wood inserted in their mutilated lips. I notice in many works of travel that, even where the men almost or entirely monopolize the ornaments, the women are always described as displaying great admiration for the beads, red cloth, and other finery taken about by travelers. I may add that I am often struck by the extraordinary folly of missionaries, who habitually preach down the love of ornament on the part of savages or of emancipated slaves (especially the women), when in reality this love is the first step in aesthetic progress, and the one possible civilizing element in their otherwise purely animal lives.[4] It ought rather to be used as a lever, by first making them take a pride in their dress, and then passing on the feeling so acquired to their children, their huts, their gardens, and their other belongings.

Such in fact has been, I believe, the actual course of our aesthetic evolution. The feelings vaguely aroused by beautiful objects in the non-practical environment become whetted and strengthened by exerercise upon ornaments and pigments, and so extend themselves with increased vividness into new channels. Art, however rude, has especially helped on this primitive progress. The appreciation for the beautiful in man's handicraft leads on to the appreciation of the corresponding beauty in natural objects. I have attempted to trace this reaction, so far as regards the sense of symmetry, in a previous number of this journal,[5] and I shall endeavor still further in the present paper to illustrate its progress in a somewhat different direction.

From delight in the beauty of ornaments to delight in the beauty of weapons or other utensils is but a step. What a man carries in his hands is almost as much a matter for personal pride as what he wears around his neck or his waist. From the very earliest ages, the material for palæolithic stone hatchets seems to have been intentionally chosen with conscious reference to beauty of color.

Among the minerals so employed were "red or other colored jasper"; "greenstone, mottled jade, and green jasper"; "quartz, agate, flint, obsidian, fibrolite, chloromelanite, aphanite, diorite, saussurite, and staurotide." The bone knife-handles and other utensils from the rock shelters of the Dordogne (of palæolithic date) are admirably carved into the forms of animals, or decorated with ornamental patterns. Indeed, both in outline and detail, most works of art of the chipped-flint period show very distinct æsthetic care, which is often marvelous when one considers the extremely rude nature of the tools in use, and the immense extra labor entailed upon the maker by any attempt at unnecessary ornamentation. The weapons of all but the very lowest existing savages show similar marks of æsthetic care. Their stone hatchets, besides being exquisitely polished, like those of the European neolithic age, are fitted in smooth wooden handles, and bound to the shaft by pretty twisted strings of red and yellow fiber. The Australian boomerangs are beautifully worked in hard wood. The staves or clubs of the Admiralty Island chiefs are wrought with the most exquisite and laborious tracery, which puts to shame our careless European woodcarving. The canoe-paddles of other Polynesian and Melanesian tribes are models of graceful and effective ornamentation. Among many savages belonging to the second rank, I find few works of art except weapons or like personal utensils on which any high degree of pains has been expended. We may therefore fairly regard this as the second human stage of aesthetic development.

Hardly superior to this second level is the love for decoration on vessels and other domestic utensils. Yet these, as being just one degree less personal than weapons, may be regarded as occupying a slightly higher stage. Calabashes and cocoanuts are almost always carved or decorated. Pottery from the very first is more or less ornamental in form, and even among very undeveloped savages is often prettily molded with lines or string-courses. Many of Dr. Schweinfurth's Central African specimens are extremely graceful; while several of the exquisitely simple prehistoric forms unearthed by Dr. Schliemann at Troy and Mycenae have been adopted as effective models for the modern artistic Vallauris ware. France itself can produce nothing more beautiful in its own kind.

Decoration of the home is one degree more disinterested than decoration of the person or personal implements. The palæolithic savages who carved the knife-handles and etched the pictures of reindeer or mammoths, in southwestern France, still lived in caves and holes of the rock. But as soon as man began to dwell in a hut, that hut began to take the impress of his growing æsthetic tastes. Swiss lake-dwellings present regular square or circular ground-plans. Esquimau snow houses are finished with as much regularity and neatness as if they were built in the most durable material. Almost all savage huts are picturesque in shape, and some are even artistic in their simple style of architecture. The rudest tribes care for little but the exterior of their dwellings, since the interior is only used as a shelter for sleeping or a retreat from wet weather, not as a place of reception. Pride in personal possessions, we must always remember, has uniformly formed the stepping-stone on which our nature has slowly risen to a higher æsthetic level. So, we find houses beginning to be ornamented internally just in proportion as they are used for purposes of display. Even our own homes usually have the drawing-and dining-rooms much more elaborately decorated and furnished than the other parts of the house. The state-apartments of halls and palaces contain all the best pictures and the handsomest mosaic tables that their owners possess.

At this stage, the governmental and ecclesiastical impetus begins to be strongly felt. From the very beginning, indeed, aesthetic products are specially the attributes of royalty and divinity. The clubs and paddles noted above are those of chiefs alone: the Hawaiian feather mantles were taboo to the royal family: the ivory scepter and the vermilion-painted face "belonged alike to the Roman god and to the Roman king." But, when we reach a state of culture at which the royal palace and the temple are widely different from the huts of the subject, we find a great æsthetic advance. Architecture is indeed a specially regal and religious art. All early buildings of any pretensions are either palaces or shrines: only at a comparatively late stage of evolution, and under an industrial régime, do handsome mansions of commoners begin to exist. Even in our own day, if we see an exceptionally large and pretentious house, we take it for granted that it is, if not a palace, at least a public building. In India, all the great architectural works are either mosques and temples or palaces and mausoleums of native or foreign rulers. In Egypt, they are either pyramids of dead kings or fanes of still earlier gods. So, too, in Mexico, Peru, Central America. The catalogue of the works of art in Solomon's temple and Solomon's house, whether authentic or not (and good authorities accept it as historical), represents at any rate the æsthetic status of the Hebrews at the date at which it was committed to writing.

The king, then, from the first surrounds himself with such natural or artistic products as add to his impressiveness and dignity. Trophies and other decorations of warlike origin, badges and costumes, paint and ointment, have been so fully treated in this connection by Mr. Herbert Spencer in his "Ceremonial Institutions" that I need not dwell upon them further here. But a few words as to later and more developed stages may not be out of place. Architecture is the central royal art, and its first object is to "beautify the house of the king." Beginning with the regal hut, it goes on to the frail and gilded palaces of China and Burmah, the house of cedar which King Solomon builded, the vast piles of brick erected by Assyrians and Babylonians in the alluvial valley of the Euphrates, the solid granite colonnades of Thebes and Memphis, the huge marble domes of Agra and Delhi, the stucco monstrosities of Mohammedan Lucknow, Sculpture first grows up as the handmaid of architecture, and begins its modern form with the bas reliefs of Egypt and Assyria, or the rock-hewn colossi of Elephanta. We still see the conjunction between royalty and these two sister arts in the beautiful Renaissance façade of the Louvre and the tasteless gilding of the Albert Memorial. Beside the ancient Nile or in the courtyards of Nineveh, we find the subjects ever the same—the king conquering his enemies; the king hunting and slaying a lion; the king driving a herd of naked captives to his capital city. Thus the aggrandizement of royalty becomes at the same time the opportunity for the exercise and development of plastic skill, while it affords models of the beautiful in art for the admiration and the æsthetic education of the subject throng.

Similarly with painting. Beginning with the rude decoration of the savage cloak and girdle, it advances to the smearing and gilding of the royal hut. Thence it progresses to the brilliant coloration of Egyptian columns and frescoes, and to all the Memphian wealth of blue, green, crimson, and gold with which so many modern restorations have made us familiar. In India, debarred from imitation by Moslem restrictions, it produces the exquisite decoration of the Taj and the Delhi palaces: in western Islam, it gives us the gorgeous Moresque tracery of the Alhambra. In its regular European development, becoming mainly ecclesiastical during the early middle ages, it reasserts its original governmental connection in the palaces of Florence and Venice, in the Vatican, in the Louvre and the Luxembourg, in Whitehall and Hampton Court, in Dresden and Munich, in modern Berlin and St. Petersburg. Sèvres and Gobelins were originally royal factories: Giotto, Michael Angelo, Raphael, Holbein, Rubens, Vandyke, all produced their masterpieces for popes or kings—Leo X, Henri IV, Charles I. Conversely, American artists have often noted the chilling effect of the want of a court upon the aesthetic susceptibilities and creativeness of their countrymen generally. Europe has, on the whole, purchased its art at the hard price of its long apprenticeship to despotism. In India, native art has steadily died out with the gradual extinction of the native courts. In Hellas and Italy it happily survived royalty because pressed into the double service of religion and of the sovereign people in its corporate capacity. What the house of Pharaoh was to Egypt, that was the house of Athene to Athens.

The gods, indeed, have done almost more for the expansion of the aesthetic faculty than even the kings. If the savage decorates the living chief and his house, how much more must he decorate and beautify the image and the house of that greater dead chief, the o-od—that ancestral ghost whom even the living chief dreads and venerates exceedingly! Hence, from the very first, while the ornaments of the king and the god are the same in kind, those of the god are the finest in degree. As the ghost gradually expands into the vaguer grandeur of the deity, his worship is surrounded with increasing magnificence. It is the temples of Heliopolis and Benares which naturally occur to our minds when we think of Egyptian or Indian architecture. It is the pyramids and mausoleums that form the initial stage of ecclesiastical buildings. All the world over, the shrines of the gods are the most splendid of all erections: only where faith is on the decline do we find the palace or the mansion outvying the cathedral and the chapel. In architecture, in sculpture, in painting, in music, the homes of the gods are the highest expression of national aesthetic feeling. Passing from the painted pillars of Karnak to the temples of Khorsabad and the mosques of Agra, we find the same care everywhere bestowed upon the service of the deities. In Hellas, we have the Parthenon and the Theseum; we have the chryselephantine statues of Phidias, and the votive tablets of Praxiteles. The marbles of Pentelicus or Paros permitted the Hellenic Aphrodite to assume a graceful and natural pose, which would have been impossible with the stiff granite limbs of a Pasht carved out from the quarries of Syene. At Rome, we have the Capitoline Jove, yielding place at last to the palace of the Divus Cæsar and to the basilica of the Christian apostle. All classical architecture, all classical sculpture, the larger part of classical painting, and no small part of classical poetry, are directly due to the influence of the old Helleno-Italian religions. And whatever little information we can gather of the æsthetic status of the Hebrews is to be derived from the story of the hangings and vessels of the tabernacle, and the molten sea, the pillars, the bases, the lavers, and the cedar ceiling of Solomon's temple. Hebrew poetry is almost without exception devotional.

In Christian times, the connection between art and religion has been even more noticeable. Our music is directly affiliated upon the Gregorian chant, and derives its notation from ecclesiastical usages. Masses and oratorios still compose its masterpieces. Our painting has come down to us from Byzantine and early Italian models, and found its home during the whole mediæval period in the great cathedrals and churches of Italy, whence it spread to the palaces of the Florentine Medici, of the Venetian doges, and of the Genoese merchant princes, and so ultimately to northwestern Europe. The whole character of pictorial art up to the Renaissance was entirely ecclesiastical and devotional. We have fed and nursed our taste upon Madonnas and Holy Families, upon Crucifixions and Assumptions, upon St. Sebastians, St. Johns, and St. Cecilias. Our architecture is based upon the Romanesque Christian church, whose rounded forms melt into the pointed arches of the Gothic cathedral. It finds its noblest expression in Pisa and Poitiers, Milan and Venice, Cologne and Chartres, Lincoln and Salisbury. And, when the classical revival comes to restore the older schools, it produces the masterpiece of its newer style in the vast dome of St. Peter's, where the four chief arts, architecture and sculpture, painting and music, all alike find their chosen home in the central point and focus of Catholic Christendom.

Nor is it only in these more notable forms that royalty and religion influence aesthetic taste. The purple and fine linen of kings' palaces; the inlaid cabinets and parquetry floors; the jade vases and painted porcelain; the Dresden statuettes and bronze candelabra; the frescoed ceiling, tapestry wall-covers, and carved wood-work—all these belong to the royal home. Even in poetry, the Queen still keeps her laureate; and the drama, originally a sort of royal specialty, is still performed at Drury Lane by "her Majesty's servants." Similarly with religion: the stained-glass window and the marble or mosaic altar; the costly vestments and sweet-perfumed incense; the fretted roof and the sculptured reredos—these in their turn belong to the worship of God. Such royal decorations and sacred ornaments react again upon the popular taste, both actively and passively. As an active effect, they give rise to and foster artistic workmanship: as a passive effect, they educate and strengthen the aesthetic faculties of the mass. Among the lower races, the æsthetic feelings have been closely linked with the sense of proprietorship: among the higher races, they gain more and more with every step in abstractness and remoteness from the personality of the individual. It was in the vast cathedrals of mediæval Europe that modern esthetic feeling received its early education.

So far we have treated little of beauty in nature: beauty in art has occupied almost our whole attention. The latter prepared the human mind for the appreciation of the former. Of the manner in which the love for art passes into the love for smaller natural objects, which exhibit minute beauty of workmanship, I have already treated elsewhere: but the taste for scenery demands a few words here. Children and early races care little for nature: it is only among the most cultivated classes of the most advanced types that the aesthetic faculty reaches this its highest and most disinterested stage. All art is at first frankly anthropinistic. Early painting, such as that of the Egyptians and Assyrians, dealt only with human and animal figures: it represented men and women, kings and queens, gods and goddesses, hunters and lions, herdsmen and cattle: but it never attempted landscape or scenery. Mediæval art in its early stages only changed its characters to saints and angels, priests and bishops. But, as it progressed from its Byzantine type, it also gradually gave more and more importance to accessories in the background, in which hills, cities, rocks, and trees, began to play a conspicuous part. At last, after the Renaissance, landscape painting became a recognized and separate branch of pictorial art, first with an admixture of figures, wild animals, or still life, but afterward in a more fully differentiated form, with all its varieties of marine, architectural, forestine, or river subjects, its waterfalls, its clouds, its rocks, its valleys, and its heather-clad hills. Even in our own day, very young people and the uncultivated classes care little for any but figure-painting: children pass over the landscapes in their picture books, and fasten at once upon the man on horseback or the boy with a top. The first object they try to draw for themselves is a human face. So, too, with literature. All primeval literary works consist of a legend, a story historical or mythical, the tale of what some man or some god has done. To the very end, novels, plays, and biography, the most human in their interest, are the favorite forms of literature. Poetry at first is all epic or narrative: lyric and descriptive verse only come in at a much later point of evolution, and are seldom thoroughly relished by any but the most cultivated. "Tell me a story," says the youngest child. "History is the most delightful of studies," says the Roman philosopher.

We may take the Homeric poems as an excellent illustration of human aesthetic feeling in this its naively anthropinistic stage. In them we find human beauty abundantly recognized and admired: Helen, for whose sake Trojans and Achaians may well contend through ten long years; Paris, on whose eyes and hair Aphrodite pours the gift of loveliness; the golden locks of Achilles, the white arms of Here, the hazel eyes of Athene, the fair cheeks of Briseis. There is much admiration, too, for works of primitive art—the golden-studded scepter, the polished silver-tipped bow of horn, the jeweled girdle of Aphrodite, the wrought figures on Achilles's shield, the embroidered pattern on the many-colored peplum which Theanô offers on the knees of Athene. The palaces of Priam and of the Phæacians excite the warmest praise of the rhapsodist. But of scenery there is little said, as is also the case in the Hebrew poets. The garden of Alcinous is, after all, but a well-ordered fruit-orchard. Nature is only alluded to as a difficulty to be overcome by man—the barren, harvestless sea; the high, impassable mountains; the forests where roam the savage wild beasts. In the Periclean age, we have a higher but still not a very exalted standard as regards natural beauty; the "Bacchæ" of Euripides being the high-water mark of Athenian love for the picturesque, and standing out in this respect as a solitary example among its contemporaries. With the greater security of Roman rule, life became less confined to the immediate neighborhood of cities; mountains and forests and waterfalls became more easy to visit; and in the "Georgics" we see the result of the change. Yet even in the "Georgics" the view of nature is still very anthropinistic, and the feeling for scenery decidedly urban. What should Ave say of a poet nowadays who should apostrophize the beauties of an Italian lake "Fluctibus et fremitu assurgens, Benace, marino"? Would he not seem in our eyes to have missed entirely the whole spirit of the scene? The words might do for Huron or Ontario, but fancy applying them to Como or Garda! Nevertheless, the Roman mind had decidedly advanced in the love of nature. The Alps were still to Juvenal mere masses of snow barring the way from Gaul to Italy; the ocean was still to Tacitus a boundless waste of western waters; but the falls of Tivoli, the little fountain-head of Bandusia, the sweeping coast-line of Baiæ, the beetling crags of Terraciha, the deep volcanic basin of the Alban lake—all these could rouse aesthetic admiration and delight in the eyes of a Horace, a Virgil, or a Claudian. With the recession of the middle ages, when men were again confined to the narrow limits of towns, aesthetic feeling went back once more to the naïve anthropinism of an earlier age; but, since the Renaissance, the love of scenery has grown perpetually, and it now probably reaches the farthest development that it has ever yet attained.

But we must never forget that the taste for scenery on a large scale is confined to comparatively few races, and comparatively few persons among them. Thus, to the Chinese, according to Captain Gill, in spite of their high artistic skill, "the beauties of nature have no charm, and in the most lovely scenery the houses are so placed that no enjoyment can be derived from it." The Hindoos, "though devoted to art, care but little, if at all, for landscape or natural beauty." The Russians "run through Europe with their carriage-windows shut." Even the Americans in many cases seem to care little for wild or beautiful scenery: they are more attracted by smiling landscape gardening, and, as it seems to us, flat or dull cultivation. I have heard an American just arrived in Europe go into unfeigned ecstasies over the fields and hedges in the flattest part of the Midlands.

The reason for this slow development may be briefly traced. The minor component elements of scenery must always have been to a great extent beautiful on their own account even to children and savages. Thus, the same bright color which gave attractiveness to flowers and gems must also have given it, though more vaguely, to the rainbow and the sunset clouds, which could not similarly be utilized for purposes of ornament. Color must also always have formed an element of beauty in blue sky and sea, red-sandstone cliffs, white chalk, green meadows, and golden corn-fields. All these objects, however, being comparatively remote from personal interest, would be little regarded by the primitive mind. But, when cultivation began, the care of the husbandman and the aesthetic interest aroused by his regular neatness would naturally set up a new feeling. Straight rows of vines or olives, trim meadows, well-kept hedges, level fields of corn, excite the farmer's admiration. This is about the level ordinarily reached (though often surpassed) by the "Georgics." In the "Iliad," when a place is mentioned with any allusion to scenery, it is generally because it is "fertile," "horse-feeding," or "rich in corn"; with Virgil, it is the careful tillage of Italian peasants that provokes attention. But wild hills and rocks are mere barren, good-for-nothing wastes to the agricultural eye. A few days before writing this paper I was wandering among the beautiful wooded heights of the Maurettes near Hyères, when I came across a party of peasants taking their lunch on a little plateau outside their cottage. Wishing to apologize for my intrusion, I said a few words about the singularly lovely view which their house commanded across the mountains and the sea. "Ah, yes," said one of the peasants in his Provençal patois, "there isn't much to see this way except the forest; but down there," pointing behind him in the opposite direction, toward the great cabbage-garden which covers the alluvial plain of Hyères—"down there one sees a magnificent country." The one view was like a bit of miniature Switzerland; the other, like a huge market-garden, as flat as this page.

Even in our own time and place, among our own race, one may see a similar æsthetic level with farmers and laborers. "So you're going to Devonshire," said a Lincolnshire yeoman to his minister (from whom I have the story); "you'll find it a poor sort of country after this. You'll never see a field of corn like ours down there, I take it." "Your country, sir," says a distinguished American visitor in England, "is very beautiful. In many parts you may go for miles together, and never see a tree except in a hedge. Nothing more beautiful can be conceived." (I take the words down from the report of an "interviewer.") To the farmer, hills like those of Devonshire were mere obstructions to ploughing: in the eyes of the practical American, trees were simply objects to be stumped and annihilated in the interest of good farming.

So long as communications are difficult and roads bad, this agricultural aspect of natural beauty will remain uppermost. It is difficult to appreciate scenery in the midst of practical discomforts. The Alps were naturally mere barriers of snow to Hannibal and Cæsar. The Scotch Highlands were less beautiful to Lowlanders when they were inhabited by hostile clansmen with a taste for cattle-lifting. Even in the last century, one is struck by the many serious discomforts which Johnson suffered in going to the Hebrides or traveling through Wales. Telford's Holyhead road must have done much to quicken the æsthetic sensibilities of the eighteenth century in England. I have myself noted in Jamaica how much the appreciation of really beautiful scenery is spoiled by the discomforts of the climate and the difficulties of transport. In such circumstances, an æsthetic feeling for scenery can hardly develop itself. Still less could it do so during the perpetual state of siege in the middle ages, or the constant warfare of the little Hellenic republics, when no man could travel a few miles from home save on urgent business and with due precautions. A lovely pass or a frowning gorge can hardly become beautiful in the eyes of those who see in it everywhere a lurking brigand.

On the other hand, when traveling becomes easier, a taste for scenery naturally arises. All the mental elements of the taste are already present; only their combination is wanted to complete the aesthetic growth. Tastes educated and refined by the arts of the city must find beauty ready to hand in much of the country. The garden and park, the Italian terrace and the Versailles avenues, the ornamental grounds and artificial lakes of the last century, formal as they seem to us now, show the gradual growth of the taste. A view from the castle or the hall becomes a desideratum. To look out upon fresh green fields and trees rather than upon the walls and narrow streets of a city must always have been pleasant to all but the most restrictedly anthropinistic minds—though even in our own day there are many townsmen who would find more to interest them in a crowd of people than in the loveliest scenery on earth. Again, only highly cultivated minds can thoroughly enjoy the beauty of places which have been always familiar from childhood: and we can hardly expect a taste for scenery to develop among people who necessarily live (like all but the most civilized) in one narrow place for all their days. Under such circumstances, the perception of its beauty can never arise. The habit of making tours, at first confined to the very wealthy, but gradually spreading down to the middle classes and the mass, has undoubtedly had an immense effect in strengthening the love of nature. Those who only know the stereotyped features of their own suburban fields, often flat and unlovely, cannot acquire any deep interest in scenery. But when Wales and Scotland, Auvergne and Brittany, Switzerland and the Tyrol are thrown open for us all, the habit of comparing, observing, and admiring grows upon us unawares. Those railways which Mr. Ruskin so cordially despises have probably done a thousand times more for promoting a love of beauty in nature than the most eloquent word-painting that was ever penned even by his own cunning and graceful hand.

If one may trust an individual experience, it is not the first waterfall that charms the most. Niagara itself, when seen in early youth, does not produce nearly so strong an impression as the little Swallow-Fall at Bettws-y-coed in later years. The more one sees, the more one learns what to expect, what to observe, what to admire. Here it is the wind-shaken foam-streak of the Staubbach; there, the little dancing cascades of the Giesbach; and here again, the vast unbroken emerald-green sheet of the Horseshoe Fall, pouring in ceaseless majesty into the seething turmoil of waters at its mist-begirt feet. . Each, has its own beauties of grace, prettiness, or sublimity, and each is largely apprehended and appreciated by means of half-unconscious recollections of the others. Between the American and Canadian falls at Niagara, a little belt of water forces its way through the gap which severs Goat and Luna Islands, and forms a minor cataract of its own, hardly heeded in the presence of the two great rivers plunging headlong at its side. If one fixes one's attention for a few moments on this little sheet of foam, one recognizes after a while that it is really larger than any cascade in western Europe. And, if you then turn your eyes to the vast semicircle of deep-green water on your right, you feel at once that without that standard of measurement your eye and brain would have failed adequately to grasp the mighty dimensions of Niagara.

Thus, step by step, in our own individual minds, and in the history of our race, the æsthetic faculty has slowly widened with every widening of our interests and affections. Attaching itself at first merely to the human face and figure, it has gone on to embrace the works of man's primitive art, and then the higher products of his decorative and imitative skill. Next, seizing on the likeness between human handicraft and the works of nature, envisaged as the productions of an anthropomorphic creator, it has proceeded to the admiration for the lace-work tracery of a fern or a club-moss, the sculptured surface of an ammonite, the embossed and studded covering of a sea-urchin, the delicate fluting of a tiny shell. Lastly, it has spread itself over a wider field, with the vast expansion of human interests in the last two centuries, and has learned to love all the rocks, and hills, and seas, and clouds, of earth and heaven, for their own intrinsic loveliness. So it has progressed in unbroken order from the simple admiration of human beauty, for the sake of a deeply seated organic instinct, to the admiration of abstract beauty for its own sake alone.

1 - I should like to add parenthetically that, since the appearance of my work "The Color-Sense" and the numerous criticisms to which it gave rise, I have fully reconsidered the whole question of sexual selection in the light of all that has been written about it, and feel only the more convinced of the general truth of Mr. Darwin's views upon the subject. It may be naturally objected that I am not an impartial witness in this matter: but I should like further to state that, on examining the various authorities, pro and con, I find in every case that the persons who are uncommitted to any special theological, quasi-theological, or metaphysical theory of evolution agree in full with Mr. Darwin, while only those differ from him who are bound down, en parti pris, to some more or less supernatural view of evolution, like Mr. Wallace, Professor Mivart, and Mr. J. J. Murphy, and who are therefore averse to any naturalistic explanation of the sense of beauty. I hope hereafter and elsewhere to enter more fully into this important question.

2 - This doctrine has been admirably illustrated by Mr. Herbert Spencer so far as regards the human species, in his essay on "Personal Beauty," which, though published long before the appearance of the "Descent of Man," really contains the germ of the doctrine of sexual selection.

3 - I noticed in Jamaica that the negroes generally considered as pretty negresses the same women as we should ourselves have selected among them; and many persons who have traveled among various savage races, and whom I have had an opportunity of questioning, confirm this general conclusion.

4 - I once asked a West Indian official of great experience and liberal views whether, in his opinion, Christianity had done any practical good to the negroes; and I was much struck by his answering: "Oh, yes! It makes them dress up in good clothes once a week, and so gives them an object in life for which to work and save."

5 - See an article on "The Origin of the Sense of Symmetry," in "Mind," xv.

SIR CHARLES LYELL

Six years after Sir Charles Lyell's death, his sister-in-law, Mrs. Lyell, has given the world his letters and journals in two bulky but vastly interesting as well as really valuable volumes. The book is not exactly a biography in the ordinary sense, for the editor's part has been confined to a few stray connecting paragraphs of the baldest explanation; nor is it a deliberate autobiography, for Lyell was far too unobtrusive of his own personality to sit down and write at full length about himself; but it is unconsciously autobiographical for all that, consisting of letters extending over more than half a century, and enabling us to trace in minute detail the gradual unfolding of their writer's ideas. As a study in psychological evolution these volumes are invaluable; they set before us vividly the prior causes which produced Lyell, the environment which affected him, and the influences which molded or developed his inherent faculties. Their interest is thus rather social and personal than merely geological; it is Lyell the man, not simply Lyell the writer, that they paint for us with such graphic fidelity.

Whence did he come? What conditions went to beget him? From what stocks were his qualities derived, and why? These are the questions that must henceforth always be first asked when we have to deal with the life of any great man. For we have now learned that a great man is no unaccountable accident, no chance result of a toss-up on the part of Nature, but simply the highest outcome and final efflorescence of many long ancestral lines, converging at last toward a single happy combination. Whatever he possesses he has derived in the main from his ancestors, though he may possibly add some few elements himself by functional use; and it is not, perhaps, too much to say that the most richly endowed natures must necessarily derive many of their separate endowments from very different preceding strains. In Lyell's case the ancestral facts are clear and simple. His father was a Lowland Scotch laird, a man of cultivation and refinement, with tastes wide enough to embrace both literature and science. He was a botanist of some distinction, of whose researches into the cryptogams Humboldt himself spoke with favor; and later in life he became an enthusiastic Dante scholar, collecting every known edition, and publishing numerous translations from the Florentine poet. Thus the father already foreshadowed the special combination of tastes to be found in the son. His mother came from a good Yorkshire family—the Smiths of Maker Hall, in Swaledale—and we can well believe Mrs. Lyell's statement that she was a woman of sound sense, for all her children seem to have inherited more than their father's share of intellect and vigor. Charles was the eldest of ten, having two brothers and seven sisters. All were able, but he was the ablest. The firstborn of a wealthy and cultivated family, with ample means and ample leisure, endowed by nature with literary and scientific potentialities, brought up in the stimulating atmosphere of his own home, of Oxford, and of the London literary world, surrounded from his childhood upward by men of science and men of letters, it would have been strange if Charles Lyell had not turned out exactly such a man as we all know him to have been. He was predestined for his work by the inevitable forces of his own constitution and the environing society, and he was admirably fitted beforehand for the work he had to do.

"Unencumbered research," as Mr. Sorby calls it, is, in fact, the key-note of Lyell's history. Like most other of our greatest scientific generalizers, he was brought up in an easy position, which enabled him to devote his life to science alone, without troubling his brain about the often absorbing question of the bread-supply, that wastes the best years of so many lives fit for better things. He came to us from the eighteenth century. Charles Lyell was born at Kinnordy, in Forfarshire, his father's estate, on November 14, 1797. But the real home of his childhood was Bartley Lodge, in the New Forest, which his father leased for twenty-eight years shortly after Charles's birth, though the family often returned for a time to Kinnordy as their summer quarters. The fragment of early autobiography which Lyell wrote years after for his future wife gives us some pleasant glimpses of the boy's life among the big trees and shady avenues of the Hampshire woodland. He felt the charm of nature and the open air from his childhood upward. He knew every clump and every single tree in the park, and to each one he gave a separate name. At Old Sarum, whither he used to go on half-holidays from his school at Salisbury, he loved already to break the flints from the chalk to see which had crystals of chalcedony in the middle, and which had white cores of sparkling quartz. Even then, before he was eleven years old, he had taken to collecting beetles and butterflies, finding out their names from the entomological books in his father's library. This free life in the New Forest must have formed such a preparation for his future work as could have fallen to the lot of very few boys in England; nowhere else, perhaps, in this over-tilled kingdom could he have formed so just an idea of what Nature left to herself is like— though even the New Forest looks but an artificial thing, after all, beside genuine native primeval woodlands. Moreover, he luckily escaped the conventionalizing and stereotyping drill of our public schools; he was never put through one of those dismal mills for crushing out individuality, into which we turn most of our best material, so as to grind it down to the Procrustean measure of Ovidian elegiacs and Æschylean trimeters. He went to three small private schools, first at Ringwood (close to home), then at Salisbury (where "we bad the very best boys in Wilts, Dorset, and Hants"—a touch of a sort that dies out of his letters or journals with the course of time), and finally at Midhurst, in the very heart of the Weald of Sussex. He was thus spared the brutal influence of "compulsory football," which would have been substituted for the pursuit of nature in a modern public school. His tutors, indeed, shook their heads at his solitary ways, but they only gently hinted that they were unmanly. Our enlightened modern head masters would have severely reprimanded him for "loafing."

On the other hand, the boy's school-training laid the foundation for that wide and general culture which was afterward so markedly to distinguish him, even among the cultivated scientific men of his own time. The danger of becoming a narrow specialist, with no eye for anything on earth except the last rare thing in ammonites, was obviated in great part by the direction given to his natural tastes at Midhurst. He "had a livelier sense than most of the boys of the beauty of English poetry," he tells his wife, long after. "Milton, Thomson, and Gray were my favorites, and even Virgil and Ovid gave me some real pleasure, and I knew the most poetic passages in them." Scott dazzled his boyish fancy as he dazzled all the world while the present century was in its teens; and when a school competition was proposed for the best English verse in the ordinary heroic decasyllabic couplets, Lyell Senior boldly sent in his copy in the metre of the "Lady of the Lake," and won the prize, too, in spite of innovation attempted and rules openly infringed. Some burlesque Latin hexameters which he wrote about the same time lingered in his memory till past middle life—an epic suggested by the Batrachomyomachia, and devoted to the draining of the play-ground pond, much infested by predaceous water-rats. Such things are small in themselves, no doubt; every promising lad of literary tendencies at every big school has done the very same in his time, without setting the Thames on fire, after all; but they are valuable as marking the specific admixture which made Lyell something other in after-life than the mere bone-hunter or snailcatcher of scientific societies. Heaven forbid that our future geologists should all be cast in the uniform mold of the classical tripos!—but there was a certain tinge of the humane letters about these savants of the last generation which relieved them from the chilliness, the austerity, and the want of human interest that many people notice as a defect among the average scientific men of the present day.

At seventeen—young even for those days, I fancy—Lyell went up to Oxford. His college, Exeter, was still almost exclusively a westcountry one, and west-countrymen were not popular nor remarkable in the university for polished manners. He tells his father a mythical story how some Devonshire man at Exeter was asked by the examiners, "Who was Moses?" "Moses," says the examinee, "I knows nothing about Moses; but ax me about St. Paul, and there I has 'ee." Good evidence how long provincial prejudices lingered in Oxford, as they still linger about the Jesus Welshmen and the Balliol Scots. The letters from college (anno 1817) are amusingly old-fashioned in their eighteenth-century echoes. They are written stiffly in the literary style of the past generation, with Horace deliberately dragged in, thus:

"Hunc varum distortis cruribus."—Sat.

But we are gainers hereby in the end; for Lyell's epistolary style, thus developed, was very different from the hasty manner of the present day, based upon the post-card and the telegraph-form.

It was at Oxford, too, that Lyell discovered geology, hitherto to him a terra incognita, or, rather, inopinata. He attended Buckland's lectures, and seems at once to have been converted to the new love, the insects being henceforth almost entirely deserted, or, at least, relegated to the second place. One of his long vacations was spent at Yarmouth with the Dawson Turners; and already we see the theory of "causes now in action" fermenting in his eager brain. He visits the alluvial delta of the Yare, finds evidence of ancient channels blocked up by the shingle which so diverted the course of the river, learns that Norwich was a great port in mediaeval history, and, putting two and two together, comes to the natural conclusion that the changes in that part of the coast were very recent, and were due, not to one of the then fashionable cataclysms, but to river-silt still in course of deposition. "Cromer, Bakefield, Dunwich, and Aldborough," he says, "have necessarily been losing in the same proportion as Yarmouth gains." The bent was there even at this early date; and it is the bent that makes the man. The old drastic cosmogony was trembling to its fall; the germs of evolutionism were already in the air. Catastrophes, special creations, deluges, and the rest, though backed by the great name of Cuvier, had had their day. Lyell was to be one of the first to discover the cumulative value of the infinitesimal. From the first, his thoughts pointed in that direction; and though he did not know to what grand results the system was to lead us in the hands of Darwin— though, indeed, he was slow to accept the results when flashed upon him too dazzlingly at last—yet it is interesting to observe how throughout he keeps a keen eye upon all the crude theories that make in the same way, such as that of Lamarck, who from the beginning exercised an obvious fascination upon his kindred mind.

Toward these final results Lyell's own work led slowly up. Perhaps it is not too much to say that in future ages, when the origin of the great uniformitarian system of interpreting nature is looked back upon with impartial eyes, four prominent names will stand out as representative of the evolutionary movement in the judgment of posterity. The first is that of Laplace, who applied it to the origin and development of sidereal systems; the second is that of Lyell, who applied it to the origin and development of the surface of our own planet; the third is that of Darwin, who applied it to the origin and development of the phenomena of life; the fourth is that of Herbert Spencer, who applied it to the origin and development of the phenomena of mind, besides working up all the scattered elements of the system into one complete and harmonious whole. To pretend that Lyell stood up to the level of the other three would be passing the love of biographers: his work neither required nor engaged such high synthetic powers as theirs. But, without the first two, the revolution accomplished by the last two could never, perhaps, have been successfully carried out.

While at Oxford, too, general culture is not neglected. We find Lyell criticising Mr. Coleridge's new poem of "Christabel," writing some mild verses of his own on Staffa, which he had just visited with his father (better mild than none), and not quite successfully trying to take an interest in his tutor's lectures on the Ethics, where every Oxford man can surely afford him the most heart-felt sympathy. In 1818 he made a vacation tour through France, Switzerland, and Italy, observing and learning much, and interesting himself in art and society. He sees the Dranse in flood, and pores over the pictures of the Pitti Palace and the domes of Venice. Coming home, he went in for classical honors, and took a second in 1819. In after-life he evidently regretted the sort of teaching he had got at Oxford as much as most other men do; yet it left some good effects, apparent enough in all his subsequent work.

Law was to be his profession: so he went to Lincoln's Inn and made a beginning of reading. But luckily his eyes were weak, and he was sent abroad again for a trip to Rome. Here he devoted himself to the Forum Romanum and the Vatican, and left no time for geology—good education for his future work. Next, he is back in England, and down at Romney, with a friend. What luck for one of his bent: Yarmouth and Romney, the two great modern districts of England, the exact places to see geology now at work under one's very eyes! Here comes one of the jarring passages again: "The back door, opening into the farm-yard, betrays [his friend's father] to have been the farmer turned gentleman, not the gentleman turned farmer. How short and direct is the road through Eton and Oxford from the grazier on Romney Marsh to the fine gentleman!" But even here the better nature comes out on second thoughts—"or, to speak plainly, to the real gentleman in ideas, manners, and information." In the earlier letters there is a good deal of this sort of thing—talk of "good company," "my father's head livery-servant," and so forth; but we are still in the year 1822, and great allowances must be made for the son of a Scotch laird, living in the midst of the Tory society of the Regency, and hardly daring to trust his own native Liberal promptings. In politics he was Liberal from the first, though never a sound Radical; and in social matters the tone of his letters widens out steadily with time, till after his first American trip he comes back, say his friends, "ipsis Americanis Americanior."

Lyell's was a life of smooth success. It is wholly wanting in anything like plot-interest, because all honors came so easily to him. In the year in which he took his degree he was made a fellow of the Linnæan and Geological Societies. In 1823 he became secretary of the latter. Already he is a fast friend with Buckland and Mantell; and his sisters are his helpers in keeping his museum and the confidantes of his scientific theories or discoveries. About this time he makes many journeys to Paris, becoming familiar not only with French as a language, but with such men as Cuvier, Humboldt, Brongniart, and Constant Prevost. He mixes in all the best salons of that shameful period. Some of his letters are guarded, lest he should be "treated like Bowring, with the Bastile"; but, when he gets a chance of sending a sheet or two otherwise than by post, his pictures of the faithless, cynical, bigoted, irreligious Paris of the Restoration are vivid and graphic in every line. Humboldt confides to him his notions about Cuvier, who has dabbled in "the dirty pool of politics":

His soirées are mostly attended by English (says Humboldt); the truth is the French savants have in general cut him. His continual changing over to each new party that came into power at length disgusted almost all, and you know it has been long a charge against men of science that they were pliant tools in the hands of princes and ministers, and might be turned which way they pleased. That such a man as Cuvier should have given a sanction to such an accusation was felt by all as a deep wound to the whole body. And what on earth was Cuvier to gain by intermeddling with politics?. . . You well know with what contempt the old aristocracy of all countries are apt to regard all new men of whatever abilities. We feel that but too much in Germany; but here it is a principle of party to carry such prejudices to the utmost length. Cuvier's situation was a proud one while he stood in the very foremost rank of men of science in France; but when he betrayed the weakness of coveting ribbons, crosses, titles, and court favor, he fell down to the lowest among his new competitors. However, after saying so much at second-hand, Lyell adds his own opinion that Cuvier is more liberal and independent than most Frenchmen. He dares to speak well of Napoleon, the sun that has set: We must not forget (he says) that Baron Humboldt and he are the two great rivals in science, for Laplace and the mathematicians do not come in contact with them. Humboldt's birth places him on the vantage-ground; and Cuvier perhaps tries to compensate this by a little political power. As for his ratting so often, defendit numerus; what French politician could throw the first stone at him? Humboldt's family is noble and ancient in Germany; his elder brother a man now in great power there. His talents entitle him to regard with the contempt which he expresses, and I have no doubt feels, mere rank; but we may say of him, as Chateaubriand said of our English peers, that he is well aware that, while he gets too liberal, he is in no danger of losing the station and the advantages which his birth insures for him.

The young English visitor saw all that was worth seeing in this profoundly rotten society. Making every allowance for good introductions and a less crowded stage of European life than ours, the ease with which he got to know everybody seems nowadays almost incredible. At the door of the observatory he meets Laplace, "a very fine-looking old gentleman"; and he is shown over the building by Arago in person. Madame Pichon, a famous beauty, who sat for Gerard's "Psyche," admits him to her salon. Férussac shows him all his snail-shells, and tells him some things about geology that he did not know before, together with many baseless theories, which his good sense cavalierly rejects. He sees something of the intriguing great world, too; some of the chameleon-colored politicians, the scheming abbés, the fashionable Ultras, and the still more fashionable Ne Plus Ultras, as he once calls them. "Every other man one meets is either minister or ex-minister. They are scattered as thick as the leaves in autumn, stratum above stratum." He is full of interest, too, in social and political questions; writes with acuteness anent the system of subdividing the land, discusses the centralizing tendency introduced by Napoleon, and is keen about the pensions bestowed on Pairs de France by the Bourbons durante bene placito—a gift which, he says, neither blesses him who gives nor him who takes it. As yet he has done nothing serious in the way of book-making; but who would exchange such preliminary training as this for the very best and carefullest field drudgery of the mere cut-and-dried technical geologist?

However, he was not idle all this time. On the contrary, he was running up and down and to and fro upon the face of the earth, inspecting its crust everywhere, with an eye to future results; and to run to and fro was of course a far more difficult thing in the twenties than it is in these later days of easy locomotion. His letters are full of his observations taken in on the spot. Now he is down in the Isle of Wight, examining the cliffs from Compton Chine to Brook, and surprised at the careless way Buckland "galloped over the ground"—"he would have entirely overlooked the Weald clay if I had not taken him back to see it" (clearly what satisfies the Bridgwater treatises and the dean in the way of research will not satisfy this very heterodox young man); now he is investigating the tertiaries of the Paris basin at Bas Meudon; and now again he is down at Lyme Regis, classic land of geologists, watching Mary Anning, the self taught fossil-finder, unearthing the skeleton of a "superb ichthyosaurus." Every letter almost teems with new facts or discoveries; and Lyell's ears are open for everything new in the geological line from the ends, of the earth inward.

In 1825 his eyes had so far recovered that he was called to the bar, and went the Western Circuit for two years. He was but a dabbler at the law, however, and fortunately never gave up to the Queen's Bench what was meant for mankind. In 1826 he was elected to the Royal Society, ætatis twenty-nine. A year later, his review of Scrope's book on Auvergne, in the "Quarterly," clearly showed the line that he meant henceforth to adopt. He came forward as the champion of the views set forward by Hutton and Playfair—views which he was to modify profoundly, to make his own, and to stamp with the seal of universal scientific recognition. About this time he conceived the plan of the "Principles of Geology," his first epoch-making book. Shortly after, he went abroad with Murchison to France and Italy, collecting material for the great work. His letters home bristle with amusing sketches of his Sicilian experiences, for Sicily was then even more impassable off the grand route than it is now; and he often had to rough it in strange quarters. He has a keen eye for the ludicrous side of things, and tells many odd stories of men and manners. "This, signor," says his cicerone once, "is the wife of Pompey the Great, named after Pompeii; she is weeping her husband's death, who was killed at the siege of Troy." At Girgenti he sees "a droll sight. Fifteen orphan boys were paraded before the statue stark naked on a windy day, and then clothed by the bishop in the name of the king." He has time, too, besides climbing Etna, and noticing such things as the signs of the rise and fall on the famous temple at Pæstum, to look at Giotto's frescoes, and to observe much about men and politics. At the end of his tour he writes from Naples to Murchison (who had not accompanied him so far):

My work is in part written, and all planned. It will not pretend to give even an abstract of all that is known in geology, but it will endeavor to establish the principle of reasoning in the science;. . . that no causes whatever have, from the earliest time to which we can look back, to the present, ever acted, but those now acting; and that they never acted with different degrees of energy from that which they now exert. I must go to Germany and learn German geology and the language, after this work is published, and before I launch out into my tables of equivalents. . . . This year we have by our joint tour fathomed the depth and ascertained the shallowness of the geologists of France and Italy as to their original observations. We can without fear measure our strength against most of those in our own land, and the question is whether Germany is stronger. They are a people who generally "drink deep or taste not." Their language must be learned; the places to which their memoirs relate, visited; and then you may see, as I may, to what extent we may indulge dreams of eminence at least as original observers.

It is a great thing that Lyell was able thus to devote himself entirely to his work, and to spare no expense or trouble that would render him more competent rightly to perform it. "I shall never hope to make money by geology," he said; and again, "I will waste no time in book-making for lucre's sake." To travel everywhere and see everything with his own eyes was his great idea: "We must preach up traveling, as Demosthenes did delivery, as the first, second, and third requisites for a modern geologist." In 1830 the first volume of the "Principles" came out, and immediately achieved a marked success. No sooner was his hook published, than he was off to the Pyrenees, and dashing down in his impetuous way into Catalonia. Here he mixes up in his letters the volcanoes of Olot and the salt mines of Cardona with much amusing chat about the peninsularity of the Spaniards and the odd people he met en route. On his way back through France, he comes across the tail-end of the Revolution of 1830. At Perpignan he sees the cross removed from the cathedral, and hears a bystander indulge in the exquisitely French reflection: "Chacun a son tour; le bon Dieu a eu le sien." Next year he is off to Germany, inspecting the volcanic region of the Eifel. About the same time he accepts the professorship of Geology in King's College, offered him by three bishops, who knew not what they did; for Conybeare vouched for his orthodoxy. Even then Conybeare must have been satisfied with very little. Lyell did not keep the chair, however, as it interfered with his schemes of traveling and original research. So he returned immediately to his tours, much to the ultimate advantage of science, and no doubt to the great satisfaction of the hesitating episcopal triumvirate.

During all these bachelor years Lyell was daily mixing with the most cultivated society of the time. In every letter half a dozen well known names catch the eye at once. On one page, he is dining at Craig Crook Castle with Francis Jeffrey, "a great treat," and meeting "Mr. Maculloch, who gave the celebrated lectures on political economy in town last summer, which I attended"; on another, he is breakfasting at Lockhart's with Sir Walter Scott, "a far more genteel-looking man than Phillips has represented him in his portrait"; and on a third, he is at Cambridge, playing whist with Copley, Master of the Rolls, afterward Lord Lyndhurst, and chronicling only "a stiff bow" from highly-aristocratic young Lord Palmerston, who must then have been strangely different from his later easy-going self. Mrs. Somerville was always a close friend, and he even chaperones her to a Sunday evening "At Home" at Sir George Phillips's, where they meet Yankee novelist Cooper, politico-economical Mrs. Marcet, ethical Mackintosh, poet Rogers, Benthamite Dumont, Conversation Sharp, Sir Walter himself, and a dozen other assorted notabilities. Sir John Herschel, too, was an equally early ally, to whom many of the letters are addressed. Lyell is very catholic. He goes to hear Paganini, not enthusiastically; and then he goes to kirk to hear Chalmers, and retains enough of the Scotchman about him to characterize the sermon as "a very long discourse, but admirable." This catholicity comes out in far stronger relief in his letters than even in his published works, which stick comparatively close to the matter in hand. One sees it over and over again in such little touches as his first notion that he might write the "Principles" as conversations on geology, in the form of "a dialogue like Berkeley's 'Alciphron,' between equals." How many geologists of the new school have ever heard of the 'Alciphron,' or even know Berkeley in any other way than through one eternal quotation from "Don Juan"?

In 1831 the journal written for his future wife begins, so that we may conclude he was then or thereabout first engaged. In 1832 he married. His wife was a daughter of Leonard Horner, and a lady of tastes very similar to his own. Perhaps one may hint that all the ladies of Lyell's family were a trifle more learned than all the world would care for: it must have been rather a strain to live up to such a constant stimulation in the home circle; and most men would hardly wish to fill their letters to their wives with highly interesting details of dip, strike, and horizon. But this is a matter of personal taste. Lyell seems to have been one of the giants who can stand such incessant high pressure; and he was probably all the happier for his well assorted marriage. He himself seems strongly to have believed that bachelorhood was not good for the cause of science.

The summer of 1834 was spent in Scandinavia. Lyell was delighted with all that he saw in this new field. "There is much doing here which is unknown in England and France," he writes from Copenhagen. "I am more than ever struck with the extreme slowness with which science travels, what with multiplicity of languages, douanes, etc." If even Lyell felt this, though he spoke English, French, and Italian fluently, German well, and Spanish a little, how much must it stand in the way of lesser people, with smaller means and narrower accomplishments! After seeing Denmark from top to bottom, he crossed to Malmö and Lund, and did the Peninsula pretty thoroughly. At Stockholm, Berzelius took him in hand and gave him the cream of all he knew; at Upsala, it seems a strange link with the infancy of science to read that the daughters of the great Linnæus himself showed him over their father's garden. Conversation was limited to German, eked out, when needful, with Latin, which Lyell often found of service as a lingua franca in out-of-the-way places; but educated Scandinavians usually speak English so well that even the most helpless foreigner is seldom at a loss. He seems to have been as pleased with the peaceful and simple descendants of the wickings as most other people, and to have returned to Scandinavia with special pleasure on future visits. In 1837 he took his wife with him, and made further investigations on the geology of the Baltic basin, which stood him in good stead in his later works.

Naturally, as he grew older, after the "Principles" and the "Elements" had made their mark, he became an authority, and saw even more of the best intellects of the time than before. His correspondence with Mr. Darwin—not yet the apostle of evolution—seems to date from this period, and the allusions to London society crowd more and more thickly on every page. The tone, however, remains unchanged. Not a trace of narrow specialism anywhere. We get long accounts of such events as a party at Milman's, where Rogers and Whewell discuss Pope, and where Milman gives the fresh opinion of a contemporary on Macaulay's "Bacon." To follow him in all his wanderings after the age of railways would be impossible: a run across to Spain, Italy, or Scandinavia, seemed to him merely an ordinary bit of his week's work. In 1841, however, he took a more ambitious trip across the Atlantic to lecture at the Lowell Institute, and then traveled through much of the United States and Canada. Geologically, he was deeply impressed by the great scale of the phenomena he saw, the vast lakes, the enormous glacial deposits, the immense subterranean forests; socially and politically, the trip left lasting effects upon his tone of mind. Singularly unprejudiced to start with, he met American society frankly and cordially, and judged both its merits and defects with somewhat lenient impartiality. But his kindliness was not the result of mere unobservant and uncritical good nature. He kept his eyes open, as usual, to all the main sociological factors, and rightly remarks that many Englishmen set down much to American political institutions which is really due to American circumstances—abundant land, free elbow-room, and constant European immigration, often of the poorest and most ignorant class. On the other hand, when he crosses the border at Niagara, he sees the weak points of the colonial system on the north of the Great Lakes keenly and acutely:

You and I would hear more in good society here (in Canada) in one week (he writes to Leonard Horner), which we should consider narrow-minded and prejudiced and ungenerous to foreigners, in matters of politics, religion, and political economy, than we heard in nine months in the United

States; for they have here all the Kleinstädterei of a colony and the enmity of the borderer, added to everything that you might disapprove of which they bring from home.

This is less true now than it was then, but there is still much truth in it; and it is painful to think that we have condemned Canada to such a poor and petty mock-national existence for forty years longer, since Lyell wrote, merely for the sake of our own meaningless imperial claim, which nobody ever seriously means to assert, but which everybody pretends to believe is vastly important. The interesting thing to note here, however, is the fact that Lyell should have come to so definite and just a conclusion after only a few weeks' sojourn in a new country. It is one of the many proofs of his keen practical penetration which lie scattered over every page of his memoirs and journals.

Perhaps the chief visible results of this first American trip was the formation of a close friendship with Mr. Ticknor, of Boston—a member of the well-known publishing firm—to whom many of his letters are henceforth addressed. They are among the most interesting he ever wrote, containing expressions of broad general opinions, which would hardly be needed in writing to European friends. Some of them are very characteristic at once of his wide tolerance and his marked tendency toward conciliation and compromise. For example, he writes once:

The time may be nearer than some think, when we shall have all sects endowed, which I trust will happen, instead of none being so. But, at all events, I abhor the political disaffection created in Ireland, Scotland, and England by the exclusive privileges of Church of England ascendency. It is really the power which is oppressive here, and not the monarchy, nor the aristocracy. Perhaps I feel it too sensitively as a scientific man, since our Puseyites have excluded physical science from Oxford. They are wise in their generation. The abject deference to authority advocated conscientiously by them can never survive a sound philosophical education.

He made altogether four voyages to America, always with an increasing sympathy for whatever is best in American life. Slavery troubled him much. He saw that the slaves were fairly well treated; that they worked lightly, fed well, enjoyed themselves hugely, and were profoundly careless about their own condition. He thought that, "if emancipated, they would suffer very much more than they would gain," and just at first he was half disposed to palter and parley with the accursed thing. But more thinking brought him back to himself; and, when the War of Secession came, he was firm as a rock on the right side, when all English society was going steadily wrong. No political movement of his time seems ever to have interested and excited him so much.

"If the result of the struggle," he writes to Mr. Ticknor in the very thick of the war, "could be the abolition of slavery by the year 1900, it would be worth a heavy debt and many lives, at any rate when one thinks of what most wars are waged for, not but that the Union alone is worth a long fighting for." And the longest letter, I think, in the whole correspondence, is one to his friend Mr. T. Spedding, defending his faith in the North against adverse criticism—a manly, noble, outspoken letter, which by itself sufficiently stamps its writer. A few condensed extracts are well worth making:

I admit that every people have the right of rebellion or revolution whenever they are oppressed. . . . But, so far from having any just grounds of rebellion, the South had been dominant to the last in foreign and domestic politics, had always had the lion's share in the choice of Presidents and other civil appointments. . . . In short, they rebelled simply because Lincoln's election showed them that the Republican party were at last determined to resist the extension of slavery into new Territories. . . . If such men as Gladstone and Earl Russell had been only six weeks in the United States, they would never have said what they did. . . . Lincoln and his colleagues are not the sort of men that you and I would put into a Cabinet, so far as their conventional manners are concerned; . . . but, after all, are Lords Palmerston, Clarendon, and some others, men of higher principle than Lincoln, or as high?

I am intimate with men equal to any here in literary attainments and in polish of manners, and of independent fortune, in the United States, whom I used to wish to see in power instead of the coarser class into whose hands the reins of government have been placed. But these men and the majority of capitalists would, I am sure, have knocked under to the South, and the slave-owner would have made a compromise by which his institution would have been more rampant than ever. If slavery, which was more injurious to the white man than to the negro, and which to a certain extent poisoned the political institutions of the North,. . . is got rid of, it will be owing to a very extended suffrage among a class which has had much instruction, for working-men, but to whom the aristocracy of wealth and refinement were not prepared to make great sacrifices for such an object. In a man of Lyell's antecedents and position, such reasoning is both brave and unexpected. I regret to say he observes in the same letter that he would rather fight for any number of years than let Ireland be "independent," though he admits that the Irish might make out a fair case for "repeal." Like most English Liberals, he can be just and sympathetic to Venetians, Poles, Hungarians, and negroes, but cannot go quite so low as Irishmen.

So much by anticipation. A life like this is so full of real triumphs that one almost forgets to mention such a small matter as that in 1848, when at Kinnordy, "he rode over the hills by Clova and Lochna-gar to Balmoral, when he had the honor of being knighted by the Queen." He was Englishman enough to appreciate the distinction, as well as the baronetcy which followed it later on. Nor was he insensible to the blandishments of royalty: he records the doings of little princes and princesses, when he happens to meet them, a trifle too much in the style of the special correspondent, and he details his conversations with a distinguished personage somewhat more fully than their intrinsic nature really demands. But there is not much of this sort of thing: as a rule, when he mentions a man, it is because the man is worth mentioning. The life in London during the years of full maturity is even richer in reminiscences of famous people than the earlier days. Dining at Hallam's, the great subject of conversation is the vacant editorship of the "Edinburgh"—Longman closeted for hours with Macaulay, and Jeffrey strongly opposed to letting the control go from Auld Reekie. Breakfasting at Rogers's, the veteran poet tells him how he knew a boatman who used to ferry Mr. Alexander Pope across the river at Twickenham, how Chantrey once came to his house as a workman, at five shillings a day, to receive orders for some ornamental drawing-room furniture, and how he still possessed the identical table at which Addison wrote his "Spectator" papers. Now it is "Ruskin, who was secretary of our Geological Section"; now it is "a friend of mine, Huxley, who will soon take rank as one of the first naturalists we have ever produced"; and now it is "young Geikie,. . . certainly the coming geologist and writer." His eye for men was very keen, and his predictions have almost always turned out to be correct. Of Agassiz, just settling in Boston, he says: "He will be the founder of a school of zoölogy (for he has many pupils) of a high order. His enthusiasm is catching, especially when he has a good soil to work upon." Nor do his interests narrow at all with years. "I sat long before the Madonna di San Sisto to-day," he writes from Dresden, "and can feel its beauty." At Madeira, Teneriffe, the Grand Canary, and Palma, he enlarges his notions by new sub-tropical experiences. But the great scientific and philosophical revolution of the present century burst upon him, after all, half unprepared. He has long ago demolished the Mosaic cosmogony; he is deeply interested in Bishop Colenso; he has already strong views as to the antiquity of man; and yet Mr. Darwin's "Origin of Species" comes across his horizon at last almost like a thunder-clap. The truth is, he was committed to the opposite belief, and he was old for a sudden revulsion. He accepted the new creed, indeed, slowly and cautiously, but he had a struggle for it, and it cost him hard.

Lyell's attitude toward the grand theory of the origin of species by descent with modification was, indeed, in many ways a singular one; and these letters throw much light upon the evolution of his ideas with regard to it. Though his own views as to uniformitarianism and the antiquity of man might seem naturally to lead toward the acceptance of the development hypothesis—for it is much more difficult to imagine creation taking place in the midst of an ordinary physical series of events than to imagine it taking place in order to restock a world desolated by a divinely ordered cataclysm—he formally rejected the theory as broached by Lamarck, and he hesitated for some time to accept it as altered and amended by Darwin. Indeed, to the last he was but a lukewarm convert. Unless my memory misleads me, I have heard Mr. Herbert Spencer say that the true test whether a man was an evolutionist in fiber or not was to be found in the question whether he accepted evolution before Mr. Darwin had made its modus operandi intelligible. There are men who rejected the doctrine of special creation on evidence adduced; and there are men who never for a moment even entertained it as conceivable. These latter may not always have seen the πῶς of evolution, but they always saw the ὅτι. Judged by such a standard, Lyell occupies a middle position. From his earliest days he seems to have hankered after some such naturalistic explanation of life, and yet to have feared cordially to accept it. In 1827 Mantell sent him Lamarck, when he was on circuit at Dorchester. He writes back shortly after:

I devoured Lamarck en voyage, as you did Sismondi, and with equal pleasure. His theories delighted me more than any novel I ever read, and much in the same way, for they addressed themselves to the imagination, at least of geologists, who know the mighty inferences which would be deducible were they established by observations. But, though I admire even his flights, and feel none of the odium theologicum which some modern writers in this country have visited him with, I confess I read him rather as I hear an advocate on the wrong side, to know what can be made of the case in good hands. I am glad he has been courageous enough, and logical enough, to admit that his argument, if pushed as far as it must go, would prove that men may have come from the orang-outang. But, after all, what changes species may really undergo! How impossible will it be to distinguish and lay down a line beyond which some of the so-called extinct species have never passed into recent ones!

The last two sentences show how, even then, Lyell was trembling upon the brink of the truth. He had got in the thin end of the wedge; he was prepared to admit the first infinitesimal in the long series whose sum makes up at last the difference between himself and the amoeba: and yet he refused to go any further.

Time after time, for many years, we find the same thing cropping up again. The question is always before him, though he wavers much in the way he regards it. It seems to fascinate him and draw him on; even when he is fighting against it, it appeals to him as the natural complement of his other beliefs. In 1830 he writes to his sister from Paris:

This morning all my Etna shells were examined; out of sixty-three only three species not known to inhabit the Mediterranean, yet the whole volcano nearly is subsequent to them, and rests on them. They lived, on a moderate computation, one hundred thousand years ago, and, after so many generations, are quite unchanged in form. It must, therefore, have required a good time for orang-outangs to become men on Lamarckian principles.

Anyone can see the falsity of this reasoning, which would imply an absolute uniformity in the rate of change in nature everywhere. A little later, in 1836, he writes to Sir John Herschel:

In regard to the origination of new species, I am very glad to find that you think it probable that it may be carried on through the intervention of intermediate causes. . . . An insect may be made in one of its transformations to resemble a dead stick, or a leaf, or a lichen, or a stone, so as to be somewhat less easily found by its enemies; or, if this would make it too strong, an occasional variety of the species may have this advantage conferred on it; or, if this would be still too much, one sex of a certain variety. Probably there is scarcely a dash of color on the wing or body of which the choice would be quite arbitrary, or which might not affect its duration for thousands of years.

In some ways this is marvelously near Darwin; but in others it differs toto cœlo; for Lyell does not see that these variations could arise "spontaneously," that is to say, in the ordinary course of small differences of antenatal conditions; he sets them all down directly to "the Presiding Mind." Nor does he see that they might result at last in the production of new species. Indeed, the context, which I have suppressed, takes off much from the superficial air of anticipating Darwin, which the passage nakedly quoted undoubtedly bears. A year later he tells his sister—

The latest news is that two fossil monkeys have at last been found one in India, contemporary with extinct quadrupeds, but not very ancient, Pliocene, perhaps; another in the south of France, Miocene and contemporary with Paleotherium. So that, according to Lamarck's view, there may have been a great many thousand centuries for their tails to wear off, and the transformation to men to take place.

In 1854 he notes, after an evening at Mr. Darwin's, how Sir Joseph Hooker astonished him with an account of that strange orchid, Catasetum, which bears three totally distinct kinds of flower. "It will figure," he says, "in C. Darwin's book on species, with many other 'ugly facts,' as Hooker, clinging like me to the orthodox faith, calls these and other abnormal vagaries."

Two years later, speaking of the wingless beetles of Madeira, he asks, "Was it not foreseen that wings would only cause them to be blown out to sea and be drowned?"

Soon after, meeting "Huxley, Hooker, and Wollaston at Darwin's," he is constrained to ask, "After all, did we not come from an orang?" At last the "Origin of Species" comes out, and bit by bit Lyell is compelled to give in. Even then he can reconcile himself but slowly to the new creed. "I plead guilty," he writes to Sir Joseph Hooker, "to going further in my reasoning toward transmutation than in my sentiments and imagination, and perhaps for that very reason I shall lead more people on to Darwin and you than one who, being born later, like Lubbock, has comparatively little to abandon of old and long-cherished ideas, which constituted the charm to me of the theoretical part of the science in my earlier days, when I believed with Pascal in the theory, as Hallam terms it, of 'the archangel ruined.'" To Mr. Darwin himself he writes that "the descent of man from the brutes takes away much of the charm from my speculations on the past relating to such matters." In the end he comes to the conclusion, as he idiomatically puts it, that "we must go the whole orang"; for that old mode of envisaging the facts clings to him to the last. Finally, he writes: "The question of the origin of species gave me much to think of, and you may well believe that it cost me a struggle to renounce my old creed. One of Darwin's reviewers put the alternative strongly by asking whether we are to believe that man is modified mud or modified monkey."

I have illustrated this matter thus fully because it is one which very clearly shows the weak side of Lyell's intellect. With all his breadth of mind and freedom from prejudice, he was not ever one of those who really get to the very deepest bottom of things. His tendencies were all in the right direction, and his instinct inclined him always to the true solution; but he did not build himself up a set of first principles to start with, firmly based upon a philosophical foundation, and make these the fixed criteria of his judgments throughout. His was too English a mind for that. He clung to all old beliefs as long as possible; he parleyed and temporized with the enemy; he was for effecting a compromise wherever he could, a patched-up modus vivendi which had to be tinkered anew at every fresh discovery. To the very last his acceptance of evolution was but half-hearted; he never came out and gave it the right hand of friendship fearlessly; he was always making reservations and starting difficulties, although his own beliefs fell short of it in places only by an infinitesimal fraction. "No miracle and no catastrophes in the cosmical system," he seems to say from time to time; "no miracle in the evolution of our planet; no fresh creations en bloc to repeople a desolate world; but just a very tiny miracle now and then, somewhere behind the scenes—a single new species to be created at a time, very unobtrusively, in Australia perhaps or St. Helena—that is all I ask." Whereas a thoroughly logical mind, a mind of the very highest order, would have said even before Darwin: "Creation can have no possible place in the physical series of things at all. How organisms came to be, I do not yet exactly see; but I am sure they must have come to be by some merely physical process, if we could only find it out." And such a mind could not fail to jump at the Darwinian solution the moment it was once fairly presented to it.

At the same time it would be unjust to deny that Lyell possessed and retained throughout life an unusual plasticity of thought and modifiability of opinion. It was no small thing that long after his sixtieth year he should have had the courage formally to recant in print the condemnation of "transformism" in his earlier works, and to accept, however unwillingly, the theory that he had so often and so deliberately rejected.[1]

A somewhat ungenerous critic has lately declared that Lyell often shut his eyes when brought face to face with evidence adverse to his own views. These letters abound in proofs to the contrary. Twenty years before the publication of the "Origin of Species," he writes on another subject to Sir John Herschel:

I am very full of Darwin's new theory of coral islands, and have urged Whewell to make him read it at our next meeting. I must give up my volcanic crater theory forever, though it costs me a pang at first, for it accounted for so much—the annular form, the central lagoon, the sudden rising of an isolated mountain in a deep sea, all went so well with the notion of submerged, crateriform, and conical volcanoes, of the shape of South Shetland, and with an opening into which a ship could sail. . . . Yet, spite of all this, the whole theory is knocked on the head, and the annular shape and central lagoon have nothing to do with volcanoes, nor even with a crateriform bottom.

The same spirit comes out in many other places. "I am sure I have no objection" he says in one place about some disputed Old Red fish-scales, "for I would as lief start with vertebrated animals and fresh water as with a universal ocean and the simplest forms of animal life." Perfect loyalty to fact, a complete readiness to accept anything, provided it can be shown to be true, marks Lyell's procedure throughout. It is very clearly seen in the last great work of his life, the "Antiquity of Man." As a matter of taste, it is obvious that Lyell did not relish the application of evolutionism to his own species. But he found that the facts compelled him, and he gave in. No book ever published—not even the "Origin of Species" or the "Descent of Man"—did so much to shake the common belief in the origin of our race: so far as all thinking Europe was concerned, Lyell simply demolished the current cosmogonies. More than that, by incorporating in the book Professor Huxley's remarks about the Neandearthal skull and much similar matter, he advertised the new creed in the animal origin of man with all the weight of his European reputation. The last years of his life were almost wholly spent in investigating this question of antiquity. Fifty years before, when he was at Oxford, he noted the occurrence of certain "pear-shaped flints" at Norwich, which he supposed must have "owed their shapes entirely to animals"; and all through his life he had been especially interested in the glacial period and its remains, the border-land where geology merges imperceptibly into archaeology and history. But from the Darwinian era onward he turned his attention almost entirely to the question of antiquity. He inspected everywhere, and got abundant specimens from abroad, at times not without ludicrous difficulties. Dr. Falconer had procured him a fine cast of a fossil rhinoceros; at Naples the police voted it an infernal machine, and confiscated it accordingly. After a time it was restored, but the priests kept Dr. Falconer's osteological notes, which they declared to be treasonable, as no doubt they were from an ecclesiastical point of view. After some years spent in hunting palæoliths and weighing evidence (which involved some heavy field-work for so old a man, in the Bedford drift, the Liége and Maestricht caves, and so forth), the "Antiquity of Man" finally appeared in February, 1863. In three months he had sold five thousand copies, a remarkable success for such a book. It was his last great serious work. The remaining years of his life, though still actively spent, were devoted mainly to reconsideration and revision of what had been already done.

In February, 1875, his great and useful life closed quietly and worthily. In reviewing the seventy-eight years of his labors, it is impossible to avoid seeing throughout how admirably his opportunities were adapted to the work he had to do. He was the right man, to start with; but the lines also fell to him in the right places. With equal abilities, equal ardor, and equal singleness of purpose, he could not have done so much without the happy conjunction of circumstances as well. On the other hand, the lesson of his valuable life throws only into stronger relief the utter waste of powers and opportunities on the part of most other Englishmen in like positions. Ninety-nine people out of a hundred, put in Lyell's place, would have been nothing better than masters of fox-hounds or slaughterers of tame pheasants. When one thinks of the life-work performed by such men as Lyell and the great band of thinkers to which he belonged, one sees only the best side of wealth and position: one feels for a moment half inclined to thank the constitution of things as they are here in England for the chance it offers to such broad-minded and comprehensive workers as these. But then one thinks also of the extraordinary rarity of men who so make use of their opportunities, who regard their wealth as anything more than an easy means of the vulgarest personal gratification. It is lamentable to remember all the thousands of conservatories all over England in each of which, without perceptible difference to the owner, a few useful experiments might be tried, a few valuable observations made; and yet how many of them are ever used for any other purpose than to provide distorted flowers for a dinner-table, for a lady's hair, or for a fop's button-hole? We must congratulate ourselves if now and then, at rare intervals, we get a single Lyell out of all this mass of wasted humanity. After all, that result is in itself a great thing. We have always enough of narrow specialists in science, men valuable and important in their own way, though that is not the highest way; but we have never too many of the great coordinating and organizing intelligences, who take the scattered strands of scientific thought, and weave them together into one consistent and harmonious whole. Among such men as these Lyell stands well to the front, though not exactly in the very first rank.

1 - It is curious to note, however, that he never seems quite fully to have realized the immense difference between Mr. Darwin's view and Lamarck's. A priori, creation is from the first unbelievable; but, as a matter of evidence, Lamarck failed to make evolution comprehensible, while Mr. Darwin succeeded in doing so. Hence he was able to convert many who, like Lyell, were hanging back and waiting for a posteriori proofs. Yet Lyell himself never wholly recognized the difference.

HYACINTH-BULBS

If we were not so familiar with the fact, we would think there were few queerer things in nature than the mode of growth followed by this sprouting hyacinth-bulb on my mantel-piece here. It is simply stuck in a glass stand filled with water, and there, with little aid from light or sunshine, it goes through its whole development like a piece of organic clock-work, as it is, running down slowly in its own appointed course. For a bulb does not grow as an ordinary plant grows, solely by means of carbon derived from the air under the influence of sun-light. What we call its growth we ought rather to call its unfolding. It contains within itself everything that is necessary for its own vital processes. Even if I were to cover it up entirely, or put it in a warm, dark room, it would sprout and unfold itself in exactly the same way as it does here in the diffused light of my study. The leaves, it is true, would be blanched and almost colorless, but the flowers would be just as brilliantly blue as these which are now scenting the whole room with their delicious fragrance. The question is, then, how can the hyacinth thus live and grow without the apparent aid of sunlight, on which all vegetation is ultimately based?

Of course, an ordinary plant, as everybody knows, derives all its energy or motive-power from the sun. The green leaf is the organ upon which the rays act. In its cells the waves of light propagated from the sun fall upon the carbonic acid which the leaves drink in from the air, and, by their disintegrating power, liberate the oxygen while setting free the carbon, to form the fuel and food-stuff of the plant. Side by side with this operation the plant performs another, by building up the carbon thus obtained into new combinations with the hydrogen obtained from its watery sap. From these two elements the chief constituents of the vegetable tissues are made up. Now, the fact that they have been freed from the oxygen with which they are generally combined gives them energy, as the physicists call it, and, when they recombine with oxygen, this energy is again given out as heat, or motion. In burning a piece of wood or a lump of coal, we are simply causing the oxygen to recombine with these energetic vegetable substances, and the result is, that we get once more the carbonic acid and water with which we started. But we all know that such burning yields not only heat, but also visible motion. This motion is clearly seen even in the draught of an ordinary chimney, and may be much more distinctly recognized in such a machine as the steam engine.

At first sight, all this seems to have very little connection with hyacinth-bulbs. Yet, if we look a little deeper into the question, we shall see that a bulb and an engine have really a great many points in common. Let us glance first at a somewhat simpler case, that of a seed, such as a pea or a grain of wheat. Here we have a little sack of starches and albumen laid up as nutriment for a sprouting plantlet. These rich food-stuffs were elaborated in the leaves of the parent pea, or in the tall haulms of the growing corn. They were carried by the sap into the ripening fruit, and there, through one of those bits of vital mechanism which we do not yet completely understand, they were selected and laid by in the young seed. When the pea or the grain of wheat begins to germinate, under the influence of warmth and moisture, a very slow combustion really takes place. Oxygen from the air combines gradually with the food-stuffs or fuels—call them which you will—contained in the seed. Thus heat is evolved, which in some cases can be easily measured with a thermometer, and felt by the naked hand—as, for example, in the malting of barley. At the same time motion is produced; and this motion, taking place in certain regular directions, results in what we call the growth of a young plant. In different seeds this growth takes different forms, but in all alike the central mechanical principle is the same: certain cells are raised visibly above the surface of the earth, and the motive power which so raised them is the energy set free by the combination of oxygen with their starches and albumens. Of course, here, too, carbonic acid and water are the final products of the slow combustion. The whole process is closely akin to the hatching of an egg into a living chicken. But, as soon as the young plant has used up all the material laid for it by its mother, it is compelled to feed itself just as much as the chicken when it emerges from the shell. The plant does this by unfolding its leaves to the sunlight, and so begins to assimilate fresh compounds of hydrogen and carbon on its own account.

Now, it makes a great deal of difference to a sprouting seed whether it is well or ill provided with such stored-up food-stuffs. Some very small seeds have hardly any provisions to go on upon; and the seedlings of these, of course, must wither up and die if they do not catch the sunlight as soon as they have first unfolded their tiny leaflets; but other wiser plants have learned by experience to lay by plenty of starches, oils, or other useful materials in their seeds; and, wherever such a tendency has once faintly appeared, it has given such an advantage to the species where it occurred that it has been increased and developed from generation to generation through natural selection. Now, what such plants do for their offspring, the hyacinth and many others like it do for themselves. The lily family, at least in the temperate regions, seldom grows into a tree-like form; but many of them have acquired a habit which enables them to live on almost as well as trees from season to season, though their leaves die down completely with each recurring winter. If you cut open a hyacinth-bulb, or, what is simpler to experiment upon, an onion, you will find that it consists of several short abortive leaves, or thick fleshy scales. In these subterranean leaves the plant stores up the food-stuffs elaborated by its green portions during the summer; and there they lie the whole winter through, ready to send up a flowering stem early in the succeeding spring. The material in the old bulb is used in thus producing leaves and blossoms at the beginning of the second or third season; but fresh bulbs grow out anew from its side, and in these the plant once more stores up fresh material for the succeeding year's growth.

The hyacinths which we keep in glasses on our mantel-pieces represent such a reserve of three or four years' accumulation. They have purposely been prevented from flowering, in order to make them produce finer trusses of bloom when they are at length permitted to follow their own free-will. Thus the bulb contains material enough to send up leaves and blossoms from its own resources; and it will do so even if grown entirely in the dark. In that case the leaves will be pale yellow or faintly greenish, because the true green pigment, which is the active agent of digestion, can only be produced under the influence of light; whereas, the flowers will retain their proper color, because their pigment is always due to oxidation alone, and is but little dependent upon the rays of sunshine. Even if grown in an ordinary room, away from the window, the leaves seldom assume their proper deep tone of full green; they are mainly dependent on the food-stuffs laid by in the bulb, and do but little active work on their own account. After the hyacinth has flowered, the bulb is reduced to an empty and flaccid mass of watery brown scales.

Among all the lily kind, such devices for storing up useful material, either in bulbs or in the very similar organs known as corms, are extremely common. As a consequence, many of them produce unusually large and showy flowers. Even among our native English lilies we can boast of such beautiful blossoms as the fritillary, the wild hyacinth, the meadow-saffron, and the two pretty squills; while in our gardens the tiger-lilies, tulips, tuberoses, and many others belong to the same handsome bulbous group. Closely-allied families give us the bulb-bearing narcissus, daffodil, snow-drop, amaryllis, and Guernsey lily; the crocus, gladiolus, iris, and corn-flag; while the neighboring tribe of orchids, most of which have tubers, probably produce more ornamental flowers than any other family of plants in the whole world. Among a widely-different group we get other herbs which lay by rich stores of starch, or similar nutritious substances, in thickened under-ground branches, known as tubers; such, for example, are the potato and the Jerusalem artichoke. Sometimes the root itself is the store-house for the accumulated food-stuffs, as in the dahlia, the carrot, the radish, and the turnip. In all these cases, the plant obviously derives benefit from the habit which it has acquired of hiding away its reserve fund beneath the ground, where it is much less likely to be discovered and eaten by its animal foes. For it is obvious that these special reservoirs of energetic material, which the plant intends as food for its own flower or for its future offspring, are exactly those parts which animals will be likely unfairly to appropriate to their personal use. What feeds a plant will feed a squirrel, a mouse, a pig, or a man, just as well. Each requires just the same free elements, whose combination with oxygen may yield it heat and movement. Thus it happens that the parts of plants which we human beings mainly use as food-stuffs are just the organs where starch has been laid by for the plant's own domestic economy—seeds, as in the pea, bean, wheat, maize, barley, rice, or millet; tubers, as in the potato and Jerusalem artichoke; corns, as in the yam or tare; and roots, as in arrow-root, turnip, parsnip and carrot. In all these, and in many other cases, the habit first set up by Nature has been sedulously encouraged and increased by man's deliberate selection. What man thus consciously effects in a few generations, the survival of the fittest has unconsciously effected through many long previous ages of native development.

When Sir Charles Lyell's "Antiquity of Man" and Mr. Darwin's two great works first set all the world thinking about the origin of our race, there was a general tendency among scientific men and the public generally to take it for granted that the earliest known men, those whose remains we find in the river-drift, were necessarily "missing links" between the human species and its supposed anthropoid progenitors. People naturally imagined that these very ancient men must have been hairy, low-browed, semi-brutal savages, half-way in development between the gorilla and the Australian or the Bushman. Striking word-pictures painted the palæolithic hunter for us as an evolving ape; and we all acquiesced in the pictures as truthful and accurate. With the progress of discovery, however, another phase of the question has come uppermost, and anthropologists have now for some time inclined with marked distinctness to the exactly opposite view. As we examined more and more closely the relics of the cave-men, for example, it became clear that their works of art were those not merely of real human beings, but of human beings considerably in advance of many existing savages. Professor Boyd Dawkins, who knows more about the cave-men than anyone else in Britain at least, unhesitatingly states his opinion that they were in all important respects the equals of the modern Esquimaux, whom he indeed regards as their probable lineal representatives. Anyone who has closely examined the remains recovered from the French caves cannot fail largely to fall in with this view, so far at least as regards the high level of palæolithic art. In fact, it is daily becoming clear that the antiquity of man is something even deeper and more far-reaching in its implications than Lyell himself at first imagined. For while on the one hand geologists are inclining more and more to the opinion that palæolithic man was as old as or older than the last glacial period, anthropologists on the other hand are inclining more and more to the opinion that this pre-glacial and inter-glacial man was really quite as human and quite as capable of civilization as any race now living, except perhaps a few of the most cultivated European stocks. Instead of being the "missing link," our cave-man turns out to be a mere average savage, living a rude and unintelligent life, to be sure, but quite capable, so far as regards his faculties, of becoming as civilized as the Sandwich-Islanders have become within our own memory.

It is, of course, obvious that these facts may be easily turned by opponents of Darwinism into powerful arguments against the theory of man's evolution from a lower form. "Here we accept all your facts," says the defender of the fixity of species; "we allow that man has inhabited the earth for as long a period as you choose, say 200,000 years; and, when we go down to the very beginning of that period, what do we meet with? A missing link? An evolving ape? No; nothing of the sort; a man exactly the same as the man of the present day. However far back we push our researches in the past, we find either no man at all, or else the same man that we now know. Your theory of evolution is disproved by the very facts which you were wont to allege in its favor. We used at first to argue against your facts, because we did not see in what direction they really pointed: nowadays we allow them all, and we find in them the very best bulwark of our own belief."

This argument, or something very like it, has lately been employed with great effect by Dr. Mitchell, of Edinburgh, in his able and interesting work, "The Past in the Present." The Scotch archæologist there shows good grounds for supposing that the cave-men and the river-drift men were really, in faculties and potentialities, the equals of most existing savages, if not even of our own average English population. He gives excellent reasons for the belief that while we have advanced very greatly in social organization and in material comfort since that early date, we may have advanced very little, if at all, in brain-power or in potentiality of thought. There are still isolated communities in out-of-the-way parts of Scotland which use hand-made pottery of the rudest primeval type, and spin with stone whorls of the prehistoric pattern; while their works of imitative art are ruder and more unlike the originals they depict than anything ever attempted by the earliest known men. Yet these people, as Dr. Mitchell rightly observes, are fully the equals in intelligence and moral feeling of their contemporaries in the great manufacturing centers. Hence we must not confound mere material backwardness with lowness of type or intellectual deficiency. It is probable, nay, almost certain, that the ordinary cave-man was superior in ingenuity and mental power to nine out of ten among our modern savages, and quite equal to the fair run of our own laboring classes.

Nevertheless, I believe it is allowable for us frankly to admit all these facts, and yet remain evolutionists just as heartily as before. No doubt our general tendency was at first in the opposite direction, and many evolutionists will be staggered by the conclusions of Professor Dawkins and Dr. Mitchell, while others will endeavor, under the influence of false prepossessions, to dispute their facts. But modifiability of opinion is the true test of devotion to truth, and honest thinkers can hardly fail to modify their opinions on this question in accordance with the latest discoveries. After frankly and fairly facing all the difficulties of the situation, I believe we may come at last to the following conclusions, which, for clearness' sake, I will number separately: 1. The cave-men were not only true men, but men of a comparatively high type. 2. But the river-drift men, who preceded them, were men of a lower social organization, and probably of a lower physical organization as well. 3. The earliest human remains which we possess, though, on the whole, decidedly human, are yet, in some respects, of a type more brute-like than that of any existing savages. 4. They specially recall the most striking traits of the larger anthropoid apes. 5. There is no reason to suppose that these remains are those of the earliest men who inhabited the earth. G. There is good reason for believing that before the evolution of man in his present specific type, a man-like animal, belonging to the same genus, but less highly differentiated, lived in Europe. 7. From this man-like animal the existing human species is descended. 8. Analogy would lead us to suppose that the line of descent which culminates in man first diverged from the line of descent which culminates in the gorilla and the chimpanzee, about the later Eocene or early Miocene period.

In order to give such proof of these propositions as the fragmentary evidence yet admits, it will be necessary first to clear the ground of one or two common misapprehensions. And, before all, let us get rid of that strangely unscientific and unphilosophical expression, the Stone age.

Most people who have not specially studied prehistoric archæology, and many of those who have studied it, believe that the period of human life on the earth may be divided into three principal epochs, the Iron age, the Bronze age, and the Stone age; and that the last named epoch may be once more subdivided into the Palæolithic and the Neolithic ages. All the great archaeologists know, of course, that such a division would be utterly misleading; yet, in their written works, they have often used language which has led the world generally to fall, almost without exception, into the error. The division in question can only be paralleled by a division of all human history into three periods, the age of Steam, the age of Printing, and the age of Unprinted Books; the latter being subdivided into the mediaeval and the Egyptian ages. The real facts may much better be represented thus:

There are two great geological epochs in which we find remains of man. The first is that of the palæolithic or old chipped-flint weapons. The second is the modern or recent period, including the three so-called Neolithic, Bronze, and Iron ages. The first or paleolithic epoch is separated from the second or recent epoch by a vast and unknown lapse of time. We may place its date at somewhere about 200,000 years back. The remains of human origin belonging to it all occur under the conditions which we ordinarily describe as geological; they are found either in the drift deposits of our river-valleys or beneath the concreted floors of caves. They consist chiefly of rude stone weapons, in unpolished flint, chipped off by side-blows. What events caused the break in continuity between palæolithic and recent man in Europe we do not exactly know; but many of the best authorities believe that it was brought about by the coming on of the last glacial epoch (that is to say, the final cold spell of the recurrent pleistocene cycles). If these authorities are right, then at a period earlier than 200,000 years since, Europe was peopled by palæolithic men; and about 80,000 years ago these men were very gradually driven southward by the spread of the polar ice over the whole of the northern temperate zone. Be this as it may, however, we know, at any rate, that they belonged to a far earlier state of things, when the whole geographical condition of Europe differed in many respects from that which prevails at the present day.

On the other hand, recent man in Europe dates back, probably, only some twenty thousand years or so. His remains, whether of the Neolithic, the Bronze, or the Iron age, are found in tumuli still standing on the surface of the ground. Since his reappearance here, no notable changes have taken place in the face of the country. Instead of occurring in deep natural deposits or under the solid floors of primeval caves, his bones and his weapons are found in graves or mounds of recent make. The neolithic men, though they used implements of stone, polished them exquisitely by grinding and smoothing, and were in all respects, save in the use of metals and a few similar particulars, as advanced as their successors of the Bronze age. No great gap in time separates them from the bronze and iron men, as a great gap separates all three from the palæolithic cave-men and drift-men. They were probably identical with two modern races, in three successive stages of their culture; whereas, the palæolithic race is cut off utterly from the recent race by a whole unknown interval, presumably representing the time during which Northern Europe was glaciated. Accordingly, with recent man we shall have nothing to do here.

Again, I must further premise that the very question which heads this paper—who was Primitive Man?—is in itself a somewhat irrational one. For of course, if we accept the evolutionist theory at all, there never was a first man. The early undifferentiated ancestors of men and anthropoid apes slowly developed along different lines toward different specific forms; but there never was a point in the series at which one might definitely put down one's finger and say, "Here the man-like ape became a complete man." All that we can do is to decide that the ancestors of modern man at such and such a given date had progressed just so far in their way toward the existing highest type.

Professor Boyd Dawkins, in his recent work on "Early Man in Britain," and in his discourse at the last meeting of the British Association, has so clearly summed up the results of all the latest investigations as to palæolithic man that it will only be necessary here briefly to recapitulate the views he has enunciated. He divides the men of the Pleistocene period in Europe and Asia into two successive classes, the earlier or river-drift men, and the later or cave-men. The drift of the Thames, Somme, and other rivers is the earliest geological stratum in which we find unquestionable evidence of the existence of man. The evidence in point consists entirely of chipped flint instruments of the very rudest type, incomparably ruder than anything produced by the very lowest of modern savages. Man at that period was clearly a rough and perhaps almost solitary hunter, using rude triangular stone implements. Moreover, we have evidence of that homogeneous condition which betokens an early stage of evolution, in the fact that implements of precisely the same sort are found all over Europe, Asia, and Africa. The primæval hunter who chased the stag in Africa had brethren who chased the fallow deer in Spain and Italy, and others who chased the various wild beasts among the jungles of India. Over the whole Eastern hemisphere, so far as we can judge, man was then a single homogeneous race, living everywhere the same life, and producing everywhere the same rude and primitive weapons.

The drift-men were succeeded, in Northern Europe at least, by another and higher development of humanity, the cave-men. How far they may have differed physically from their predecessors of the Drift period we have no sufficient means of judging; but the analogy of other human varieties would lead us to suspect that they presented some marked signs of advance; for we know that among all existing races there is a pretty constant ratio between social development and physical peculiarities. At any rate, the cave-men were apparently far more advanced in the rudiments of culture than the drift-men, especially toward the end of the cave period, during which they made continuous advances in the arts of life. Their weapons, though still chipped (instead of being ground, like those of the neolithic Europeans and the modern savages), were more varied in shape and better worked than the rude triangular hatchets of the drift. They manufactured, in their last stage, excellent barbed harpoons of good designs. They made fish-hooks and needles of bone with some degree of finish. They employed ruddle for personal decoration, and collected fossil shells, which they drilled and strung as necklaces. Moreover, they had a remarkable talent for imitative art, producing spirited sketches on mammoth ivory or reindeer horn of various animals, living or extinct. In fact, they seem to have been in most essential particulars almost as advanced as the modern Esquimaux, with whom Professor Dawkins conjecturally identifies them.

But if Professor Dawkins means us to understand that the cavemen were physically developed to the same extent as the Esquimaux, it is necessary to accept his conclusion with great caution. It does not follow, because the Esquimaux are the nearest modern parallels of the cave-men, that the cave-men therefore resembled them closely in appearance. Several of the sketches of cave-men, cut by themselves on horn and bone, certainly show (it seems to me) that they were covered with hair over the whole body; and the hunter in the antler from the Duruthy cave has a long pointed beard and high crest of hair on the poll utterly unlike the Esquimau type. The figures are also those of a slim and long-limbed race. And when Professor Dawkins tells us that the very earliest known man was unquestionably a man and not a "missing link," it becomes a matter of importance to decide exactly what the phrase "a missing link" is held to imply.

Man differs from the anthropoid apes mainly in the immensely larger development of his brain; for the other peculiarities of his pelvis, his teeth, and the position of his head on the shoulders, are mere small adaptive points, dependent upon his upright attitude and the nature of his food. Even the lowest savage and the oldest known human skull have a brain-capacity far bigger in proportion than that of the highest apes. Now, this brain could not, of course, have been developed per saltum it must have been slowly evolved in the course of a long and special intercourse with nature. But between civilized man and his early ancestor, common to him and the anthropoid apes, there must at some time have existed every possible intermediate link. Some such links still survive in the Bushman, the Australian black fellow, and the Andaman-Islander. Other and earlier links probably became extinct at various previous periods, under the pressure of the higher varieties from time to time developed, just as these lowest savages are now in process of becoming extinct before the face of the European colonist. But we would naturally expect the men of the palæolithic period to be still a trifle more brute-like in several small particulars than any existing savages, because they were so much the nearer to the primitive common ancestor, a few of whose distinctive traits they would probably retain in a higher degree than any race now living. In short, while it would be absurd to suppose that palæolithic men were "missing links" in the sense of being exactly half-way houses between apes and Bushmen, it is yet natural to expect that they would be the last or penultimate links in a chain whose other links are many and wanting. Do we, as a matter of fact, find any such slight traces of brute-like structure in the earliest human remains which have come down to us?

In dealing with this question we have to remember in the first place that the number of quite undoubted palæolithic human bones of the earliest period is all but absolutely nil; and that even the few dubious and suspected bodily remains which we possess, presumably of that age, are for the most part mere broken fragments. Most of our palæolithic bones belong to the latest cave age, and represent a comparatively high race of savages, known as the Cro-Magnon men. Of their earlier predecessors we know but little. We have, however, two remarkable portions of skulls, one of which is almost free from suspicion, while the other, though more doubtful, is still accepted as genuine by good Continental anthropologists. Both apparently belong to the earliest age of the cave-men. The first is the celebrated jaw of La Naulette. This is a massive and prognathous bone, with enormous and projecting canine teeth; and these canine teeth, as Mr. Darwin notes, point back very clearly to a nearly anthropoid progenitor.[1] The second is the much-debated Neanderthal skull, which possesses large bosses on the forehead, strikingly suggestive of those which give the gorilla its peculiarly fierce appearance. So good an anatomist as Professor Rolleston assures us that, if these frontal sinuses had been found without the skull to which they are attached, he would have been a bold man indeed who would venture to pronounce them human. The thickness of the bones in the rest of the Neanderthal skeleton, to which Professor Schaafhausen calls attention, also approximates to the anthropoid pattern. "No other human skull," says that able anthropologist, "presents so utterly bestial a type as the Neanderthal fragment. If one cuts a female gorilla skull in the same fashion, the resemblance is truly astonishing, and we may say that the only human feature in this skull is its size." All the skulls of what De Quatrefages and Hamy call the "Canstadt race" show these same low characteristics, and "must have presented a strangely savage aspect." The other supposed relics of the earlier cave-men are either too slight, too much crushed, or too uncertain, to be of much use for purposes of argument. When we add that even the later cave-man was almost certainly hairy, like the modern Ainos, we have before us the picture of what may fairly be considered a sort of missing link, though only the last in a long chain.

Moreover, it is a most deceptive practice to speak of the cave-men as if they were a single set of people, representing a merely temporary type. As a matter of fact, the period covered by the cave remains is enormously long, and the men of one epoch must have differed widely from those of another. M. de Mortillet has actually distinguished three subdivisions of the cave period, marked by a successive improvement in the arts of working stone and bone, to which he gives the names of the Moustier epoch, the Solutré epoch, and the La Madelaine epoch, from the stations which best typify each stage of primitive culture. M. Broca has shown that, between the time when the Moustier cave was inhabited by troglodytes and the time when the La Madelaine cave was similarly inhabited, the valley of the Vézère had undergone a denudation to the depth of twenty-seven metres; while from the date of the La Madelaine cave to our own time the denudation was only four or five metres. In other words, the interval between the two epochs was far greater than the interval between the last of them and our own times.

As to the drift-men, the few bones attributed to them are so singularly and suspiciously like those of neolithic times that it seems very unsafe to build any definite conclusion upon them. Accordingly, when Professor Dawkins tells us that "the river-drift man first comes before us endowed with all human attributes, and without any signs of a closer alliance with the lower animals than is presented by the savages of to-day," I think we must venture to suspend judgment for the present. Seeing that a later skull, like that of Neanderthal, is strikingly ape-like in one most important particular, is considerably lower in general type than that of the lowest living savage, and (as Professor Huxley has shown) is rather nearer the chimpanzee than the modern European in outline, it seems hazardous to conclude on very dubious evidence that a still earlier race had skulls as well formed as those of the neolithic Iberians. The least doubtful cases are acknowledged to be identical in character with the far later Cro-Magnon remains (belonging to the latest cave age), which in itself is enough to rouse considerable suspicion. So many supposed palæolithic skeletons, like the "fossil man" of Mentone, have turned out on further examination to be neolithic or later, that it is unwise to base conclusions upon them, when those conclusions clearly run counter to the general course of evolution.

With regard to the previous history of the human race, we can only guess at it by the analogy of the other higher mammalia. But late researches have all gone to show that the general progress of mammalian development has been singularly regular. If we apply this analogy, and couple it with the other known and observed facts, we may be able still further to bridge over the gap between man and his anthropoid progenitor. As Professor Huxley remarks, "The first traces of the primordial stock whence man has proceeded need no longer be sought, by those who entertain any form of the doctrine of progressive development, in the newest tertiaries; they may be looked for in an epoch more distant from the age of the Elephas primigenius than that is from us."

The bifurcation of the European placental mammals begins in the Eocene; and it is to the Eocene that we must look for the earliest appearance of the Primates. At that period, there existed lemurs in Europe and America, of a transitional type, showing points of resemblance to the hoofed animals of the same age, the ancestors of our own horses and tapirs. The Eocene was the epoch of the first great placental mammalian population, and we know that in such early epochs each main class, when the class is assuming a dominant position, it always possesses an immense plasticity, rapidly dividing and subdividing into more and more definitely specialized types. Accordingly, it was probably as early as this period that the ancestors of the higher apes began to differentiate themselves from the ancestors of the modern lemurs. All analogy shows us that these divisions begin a long way down in time, proceed rapidly at first, and grow less rapid as the various creatures become more and more specialized, so losing their original plasticity.

In the Miocene, the specialization of the Primates must have continued very fast; for as early as the mid-Miocene strata we find in Continental Europe a large anthropoid ape, identified by good authorities as a close relation of the modern gibbons. Other apes of the same date are similarly identified as nearly allied with other living genera. Hence the question naturally arises—if the bifurcation of the Primates had already proceeded so far in the mid-Miocene period that even existing genera of higher apes had been fairly well demarkated, must not the ancestors of man have already begun to be generically distinct from the ancestors of the other anthropoids? Is it not consonant with analogy to suppose that the monkey group should have separated from the lemur group in the Eocene; that the anthropoid apes should have separated from the monkeys in the lower Miocene; and that the human genus (as distinct from the fully developed human species) should have separated from the anthropoid apes in the mid-Miocene? There seems to be good reason for this conclusion.

In mid-Miocene strata at Thenay, the Abbé Bourgeois has found certain split flints, some of them bearing traces of fire, which he believes to be of artificial origin; and in this belief he is upheld by M. de Mortillet, Dr. Hamy, MM. de Quatrefages, Worsaae, and Capellini, and other distinguished anthropologists. Specimens may be seen in the Musée de St. Germain, almost as obviously human in their workmanship as any of the St. Acheul type. M. Delaunay has similarly found a rib of an extinct manatee, which seems to have been notched or cut with a sharp instrument; and M. Ribeiro, of the Portuguese geological survey, has noted wrought flints in the Miocene deposits of the Tagus, which he exhibited in Paris in 1879. On the evidence of these and other facts M. de Mortillet pronounces in favor of what he calls Tertiary man. But as he carefully distinguishes him from Quaternary man, "l'homme de St. Acheul"—the river-drift man of Professor Dawkins—I suppose he means to imply that this species, though belonging to the same genus as ourselves, was yet so far unlike us, so little differentiated, as to be man only in the generic, not in the specific sense.

Professor Boyd Dawkins, on the other hand, argues apparently against the existence of man in any form in Miocene Europe. "There is," he says, "one important consideration which renders it highly improbable that man was then living in any part of the world. No living-species of land mammal has been met with in the Miocene fauna. Man, the most highly specialized of all creatures, had no place in a fauna which is conspicuous by the absence of all the mammalia now associated with him. . . . If we accept the evidence advanced in favor of Miocene man, it is incredible that he alone of all the mammalia living in those times in Europe should not have perished, or have changed into some other form in the long lapse of ages during which many Miocene genera and all the Miocene species have become extinct." But, if I understand M. de Mortillet aright j this is just what he means by distinguishing Tertiary from Quaternary man. Professor Dawkins argues as though the animal which split the Abbe Bourgeois's flints must either have been man or not-man; but the whole analogy of evolution would lead us to suppose that it was really a "tertium quid" or half-man; as Professor Dawkins himself suggests, a creature "intermediate between man and something else," a creature which should "bear the same relation to ourselves as the Miocene apes, such as the Mesopithecus, bear to those now living, such as the Semnopithecus"

But Professor Dawkins, who seems strangely unwilling to admit the existence of such an intermediate link, endeavors to account for the split flints of the mid-Miocene by curiously round-about ways. "Is it possible," he asks, "for the flints in question, which are very different from the palæolithic implements of the caves and river deposits, to have been chipped or the bone to have been notched without the intervention of man? If we can not assert the impossibility, we can not say that these marks prove that man was living in this remote age in the earth's history. If they be artificial, then I would suggest that they were made by one of the higher apes then living in France rather than by man. As the evidence stands at present, we have no satisfactory proof either of the existence of man in the Miocene or of any creature nearer akin to him than the anthropomorphous apes. These views agree with those of Professor Gaudry, who suggests that the chipped flints and the cut rib may have been the work of the Dryopithecus, or the great anthropoid ape, then living in France. I am, however, not aware that any of the present apes are in the habit of making stone implements or cutting bones, although they use stones for cracking nuts." And, in a foot-note, Professor Dawkins further observes: "Even if the existing apes do not now make stone implements or cut bones, it does not follow that the extinct apes were equally ignorant, because some extinct animals are known to have been more highly organized than any living members of their class." Does not this reasoning exactly remind one of that which was current when M. Boucher de Perthes first called attention to the Abbeville flints?

Now, I confess I am at a loss to comprehend why Professor Dawkins should be so anxious to escape the natural inference that these flints were split by an ancestor of man. If he were a determined opponent of evolutionism, it would be easy enough to understand his attitude; but, as he is a consistent and bold evolutionist, one can hardly guess why he should go so far out of his way to get rid of a simple conclusion. lie argues most strenuously that man was fully developed in the Pleistocene age. He cannot imagine that man reached this full development by a sudden leap or miraculous interposition. And, therefore, he might naturally conclude that an early and less differentiated ancestor of man was living in the Miocene age, and developing upward through the Pliocene times, till he reached that highly specialized specific form which he had demonstrably attained in the later Pleistocene period. Implements such as we should naturally expect a priori to be produced by such an intermediate form are actually forthcoming in the Miocene. The traces of use and marks of fire upon them seem irresistible proofs—the edges are chipped and worn exactly like those of undoubted flake-knives—while the regular repetition of their shapes is most noticeable. Yet, for some unknown reason, rather than accept the plain conclusion of M. de Mortillet, Professor Dawkins prefers to believe that they were produced by apes, and to leave man without any traceable ancestry whatsoever. Surely he does not believe that man was suddenly evolved, at a single bound, from a creature no nearer akin to him "than the anthropomorphous apes." Yet this is certainly the conclusion which most readers would draw from his facts and arguments.

It is clear that the difficulty in all these cases depends upon the too great definiteness of our words, with their hard-and-fast lines of demarkation, when applied to the gradual and changeful forms of evolving species. The very question as to the existence and character of "primitive" man thus becomes one of mere artificial and arbitrary distinctions. We try to draw a line somewhere, and wherever we draw it we must necessarily cause confusion. Let us try, then, to set forth the probable course of evolution in the line which finally culminates in civilized man, from the Eocene age upward, using so far as possible such language as will the least involve us in classificatory distinctions.

In the very first part of the Eocene age man's ancestors were very plastic and unspecialized placental mammals of the early "generalized" type. They were still so little removed from the original form, so little adapted for special habits and habitats, that they at the same time closely resembled the progenitors of the horses and the hedgehogs. But before the middle of the Eocene period this homogeneous group had begun to split up into main branches. And by the later Eocene times the particular branch to which man's ancestors belonged had reached, even in Europe, the stage of lemuroid creatures—four-handed and relatively small-brained animals, still retaining many traces of their connection with the ancestral horse-like and insectivore-like forms. These lemuroids were forestine, and, perhaps, nocturnal fruit-eaters. They lived among trees, which their hands were especially adapted for climbing.

In the lower Miocene times the lemuroids again must have split up into two main branches, that of the monkeys and of the lemurs. We find no trace of the monkeys in the remains of this age; but, as they were highly developed in the succeeding mid-Miocene period, they must have begun to be distinctly separated at least as early as this point of time. To the monkey branch, of course, the progenitors of man belonged.

By the epoch of the mid-Miocene deposits the monkey tribe had once more presumably subdivided itself into two or three minor groups, one of which was that of the anthropoid apes, while another was that of the supposed man-like animal who manufactured the earliest known split flints. The anthropoid apes remained true to the old semi-arboreal habits of the race, and retained their four hands. The man-like animal apparently took to the low-lying and open plains, perhaps hid in caves, and, though probably still in part frugivorous, eked out his livelihood by hunting. We may not unjustifiably picture him to ourselves as a tall and hairy creature, more or less erect, but with a slouching gait, black-faced and whiskered, with prominent prognathous muzzle, and large pointed canine teeth, those of each jaw fitting into an interspace in the opposite row. These teeth, as Mr. Darwin suggests, were used in the combats of the males. His forehead was no doubt low and retreating, with bony bosses underlying the shaggy eyebrows, which gave him a fierce expression, something like that of the gorilla. But already, in all likelihood, he had learned to walk habitually erect, and had begun to develop a human pelvis, as well as to carry his head more straight upon his shoulders. That some such an animal must then have existed seems to me an inevitable corollary from the general principles of evolution, and a natural inference from the analogy of other living genera. Moreover, we actually find rude works of art which occupy a position just midway between the undressed stone nut-cracker of the ape and the chipped weapons of palæolithic times. This creature, then, if he existed at all, was the real primitive man, and to apply that term to the cave-men or the drift-men is almost as absurd as to apply it to the civilized neolithic herdsmen.

The supposed Miocene ancestor of humanity must have been acquainted with the use of fire, and have been sufficiently intelligent to split rude flakes of flint. But his brain was no doubt about half-way between that of the anthropoid apes and that of the Neanderthal skull. Such an intermediate stage must have been passed through at some time or other, and the mid-Miocene is just about the time when one would naturally expect it to have existed. The fact that no bones of this man-like creature have yet been found militates very little against the argument, for in all cases the mammalian remains, which we actually possess from any particular stratum, are a mere tithe of the species which we know must have been living during the period when it was deposited. And, after all, the works of man (or of a man-like animal) are just as good evidence of his existence as his bones would be; for, as Sir John Lubbock rightly observes, the question is whether men then existed, not whether they had bones or not.

During the Pliocene period, the scent does not lie so well, and we seem to lose sight for a while of man's ancestry. Such gaps are common in the geological history, and need surprise no one, considering the necessarily fragmentary nature of the record, based as it is upon a few stray bones or bits of flint which may happen to escape destruction, and be afterward brought to light. Some cut bones, however, have actually been detected in Tuscan Pliocenes, and may possibly bear investigation. Professor Dawkins, it is true, objects that the presence of a piece of rude pottery together with the bones casts much doubt upon their authenticity. But Professor Capellini, their discoverer, now writes that Mr. Dawkins is mistaken in this particular, and that the pottery belongs to quite a different stratum from the bones. Other marked remains have been discovered in Pliocene strata elsewhere; and worked flints have been detected in the gravels of St. Prèst, which, however, are of doubtfully Pliocene age. Nevertheless, the ancestors of man must have gone on acquiring all the distinctive human features during this period, and especially gaining increased volume of brain. If we could find entire skeletons of our Miocene and Pliocene progenitors, analogy leads us to suppose that naturalists would arrange them as at least two, if not more, separate species of the genus Homo. Whether we should call them men or not is a mere matter of nomenclature; but that such links in the chain of evolution must then have existed seems to me indisputable.

In the Pleistocene period, we come at last upon undoubted traces of the existing specific man. The early Pleistocene strata show us no very certain evidence; but in the mid-Pleistocene we find the earliest indubitable flint flake, split by chipping, and very different in type from the workmanship of the supposed mid-Miocene man-like creature. In the later Pleistocene we get the well-known drift implements. Without fully accepting Professor Dawkins's argument that the drift-men were human beings of quite a modern type, one may at least admit that the remains prove them to have been really men of the actual species now living—men not much further removed from us than the Andamanese or the Digger Indians. Accordingly, we cannot suppose that they had been developed straightway from a totally inferior quadrumanous form, and reached their Pleistocene mental eminence by a leap. "The implements of the drift," says Professor Dawkins, "though they imply that their possessors were savages like the native Australians, show a considerable advance on the simple flake left behind as the only trace of man of the mid-Pleistocene age." They also show a still greater advance upon the very rude chips of the unknown mid-Miocene ancestor. Hence the progressive improvement is exactly what we should expect it to be, and we are justified, I think, in concluding that by the beginning of the Pleistocene age the evolving anthropoid had reached a point in his development where he might fairly be considered as a man and a brother. At the beginning of that age, he was probably what naturalists would recognize as specifically identical with existing man, but of a very low variety. By the mid-Pleistocene he had become an ordinary savage of an exaggerated sort, and by the age of the drift he had reached the stage of making himself moderately shapely stone implements. The river-drift man, however, as Professor Dawkins believes, has no modern direct representative—or, to put it more correctly, the whole race, even in its lowest varieties, has now quite outstripped him, certainly in culture, and probably in physique as well.

At last, we reach the age of the cave-men. By that period, man had become to a certain extent cultured. He had learned how to make finished implements of stone and bone, and to draw and carve with spirit and with a rude imitative accuracy. It is possible enough that the cave-man was the direct ancestor of the Esquimaux, and that that race has kept its early culture with but few later additions and improvements.[2] Nevertheless, it does not at all follow that in physical appearance the earlier cave-men were the equals of the Esquimaux, or, indeed, that the Esquimaux are any more nearly related to them than ourselves. They may have been darker-skinned and less highly human looking; they probably had lower foreheads, with high bosses, like the Neanderthal skull, and big canine teeth like the Naulette jaw. Even if the Esquimaux are lineally descended from the later cave-men with little change of habit or increase of culture, the mere lapse of time, aided by disuse of parts, may have done much to modify and mollify these brute-like traits. "The fact that ancient races," says Mr. Darwin, "in this and several other cases" [he is speaking of the inter-condyloid foramen, observed in so large a proportion of early skeletons], "more frequently present structures which resemble those of the lower animals than do the modern races, is interesting. One chief cause seems to be that ancient races stand somewhat nearer than modern races in the long line of descent to their remote animal-like progenitors." We must not be led away by identifications of race in too absolute a sense. We ourselves are, of course, the lineal descendants either of the cave-men or of their contemporaries in some geologically unexplored region; yet it does not follow on that account that our late Pleistocene ancestors were white-skinned people with regular Aryan features. Granting that the Esquimaux are nearer representatives of the cave-men than any other existing race (which is by no means certain), it may yet be true that the earlier cave-men themselves were black-skinned, hairy savages, with skulls and brains of the low and brutal Neanderthal pattern. The physical indications certainly go to show that they were most like the Australian savages.

With the cave-men our inquiry ceases. The next inhabitants of Europe were the comparatively modern and civilized neolithic Euskarians—a race whom we may literally describe as historical. I trust, however, that I have succeeded in pointing out the main fallacy which, as it seems to me, underlies so much of our current reasoning on "primitive man." This fallacy lies in the tacit assumption that man is a single modern species, not a tertiary genus with only one species surviving. The more we examine the structure of man and of the anthropoid apes, the more does it become clear that the differences between them are merely those of a genus or family, rather than distinctive of a separate order, or even a separate sub-order. But I suppose nobody would claim that they were merely specific; in other words, it is pretty generally acknowledged that the divergence between man and the anthropoids is greater than can be accounted for by the immediate descent of the living form from a common ancestor in the last preceding geological age. Mr. Darwin even ranks man as a separate family or sub-family. Therefore, according to all analogy, there must have been a man-like animal, or a series of man-like animals, in later, if not in earlier tertiary times; and this animal or these animals would in a systematic classification be grouped as species of the same genus with man. In the Abbé Bourgeois's mid-Miocene split flints we seem to have evidence of such an early human species; and I can conceive no reason why evolutionists should hesitate to accept the natural conclusion. To speak of palæolithic man himself—a hunter, a fisherman, a manufacturer of polished bone needles and beautiful barbed harpoons, a carver of ivory, a designer of better sketches than many among ourselves can draw—as "primitive," is clearly absurd. A long line of previous evolution must have led up to him by slow degrees. And the earliest trace of that line, in its distinctively human generic modification, we seem to get in the very simple flint implements and notched bones of Thenay and Pouancé.

1 - Since this article was sent to press, Professor Maska, of Neutitschein, has discovered a human jaw-bone, associated with pleistocene mammalian remains, in the Schipka cave (Moravia). This bone, which belonged to a very young child (as inferred from the development of the teeth), "is of very large, indeed, of colossal dimensions."

2 - I am not, however, inclined to attach much importance to the evidence of Esquimau art; or rather, that art seems to me to point in the opposite direction. After carefully comparing numerous specimens, I am convinced that the art of the cave-men is of quite a different type from that of the Esquimaux, and far higher in kind. Both, it is true, represent animals; but there the likeness stops. The Esquimaux represent them with wooden stiffness; the cave-men represent them with surprising spirit and life-like accuracy.

THE PEDIGREE OF WHEAT

Wheat ranks by origin as a degenerate and degraded lily. Such in brief is the proposition which this paper sets out to prove, and which the whole course of evolutionary botany tends every day more and more fully to confirm. By thus from the very outset placing clearly before our eyes the goal of our argument, we shall be able the better to understand as we go whither each item of the cumulative evidence is really tending. We must endeavor to start with the simplest forms of the great group of plants to which the cereals and the other grasses belong, and we must try to see by what steps this primitive type gave birth, first to the brilliantly colored lilies, next to the degraded rushes and sedges, and then to the still more degenerate grasses, from one or other of whose richer grains man has finally developed his wheat, his rice, his millet, and his barley. We shall thus trace throughout the whole pedigree of wheat from the time when its ancestors first diverged from the common stock of the lilies and the water-plantains, to the time when savage man found it growing wild among the untilled plains of prehistoric Asia, and took it under his special protection in the little garden-plots around his wattled hut, whence it has gradually altered under his constant selection into the golden grain that now covers half the lowland tilth of Europe and America. There is no page in botanical history more full of genuine romance than this; and there is no page in which the evidence is clearer or more convincing for those who will take the easy trouble to read it aright.

The fixed point from which we start is the primitive and undifferentiated ancestral flowering plant. Into the previous history of the line from which the cereals are ultimately descended, I do not propose here to enter. It must suffice for our present purpose to say dogmatically that the flowering plants as a whole derive their origin from a still earlier flowerless stock, akin in many points to the ferns and the club-mosses, but differing from them in the relatively important part borne in its economy by the mechanism for cross-fertilization. The earliest flowering plant of the great monocotyledonous division (the only one with which we shall here have anything to do) started apparently by possessing a very simple and inconspicuous blossom, with a central row of three ovaries, surrounded by two or more rows of three stamens each, without any colored petals or other ornamental adjuncts of any sort. I need hardly explain even to the unbotanical reader at the present day that the ovaries contain the embryo seeds, and that they only swell into fertile fruits after they have been duly impregnated by pollen from the stamens, preferably those of another plant, or at least of another blossom on the same stem. Seeds fertilized by pollen from their own flower, as Mr. Darwin has shown, produce relatively weak and sickly seedlings; seeds fertilized by pollen from a sister plant of the same species produce relatively strong and hearty seedlings. The two cases are exactly analogous to the effects of breeding in and in or of an infusion of fresh blood among races of men and animals. Hence it naturally happens that those plants whose organization in any way favors the ready transference of pollen from one flower to another gain an advantage in the struggle for existence, and so tend on the average to thrive and to survive; while those plants whose organization renders such transference difficult or impossible stand at a constant disadvantage in the race for life, and are liable to fall behind in the contest, or at least to survive only in the most unfavorable and least occupied parts of the vegetal economy. Familiar as this principle has now become to all scientific biologists, it is yet so absolutely necessary for the comprehension of the present question, whose key-note it forms, that I shall make no apology for thus once more stating it at the outset as the general law which must guide us through all the intricacies of the development of wheat.

Our primitive ancestral lily, not yet a lily or anything else namable in our existing terms, had thus, to start with, one triple set of ovaries, and about three triple sets of pollen-bearing stamens; and to the very end this triple arrangement may be traced under more or less difficult disguises in every one of its numerous modern descendants. No single survivor, however, now represents for us this earliest ideal stage; we can only infer its existence from the diverse forms assumed by its various divergent modifications at the present day, all of which show many signs of being ultimately derived from some such primordial and simple ancestor. The first step in advance consisted in the acquisition of petals, which are now possessed in a more or less rudimentary shape by all the tribe of trinary flowers, or at least, if quite absent, are shown to have been once present by intermediate links or by abortive rudiments. There are even now flowers of this class which do not at present possess any observable petals at all; but these can be shown (as we shall see hereafter) not to be unaltered descendants of the prime type, but on the contrary to be very degraded and profoundly modified forms, derived from later petal-bearing ancestors, and still connected with their petal-bearing allies by all stages of intervening degeneracy. The original petalless lily has long since died out before the fierce competition of its own more advanced descendants; and the existing petalless reeds or cuckoo-pints, as well as the apparently petalless wheats and grasses, are special adaptive forms of the newer petal-bearing rushes and lilies.

The origin of the colored petals is almost certainly due to the selective action of primeval insects. The soft pollen, and perhaps, too, the slight natural exudations around the early flowers, afforded food to the ancestral creatures not then fully developed into anything that we could distinctively call a bee or a butterfly. But, as the insects flew about from one head to another in search of such food, they carried small quantities of pollen with them from flower to flower. This pollen, brushed from their bodies on to the sensitive surface of the ovaries, fertilized the embryo seeds, and so gave the fortunate plants which happened to attract the insects all the benefits of a salutary cross. Accordingly, the more the flowers succeeded in attracting the eyes of their winged guests, the better were they likely to succeed in the struggle for existence. In some cases, the outer row of stamens appears to have become flattened and petal-like, as still often happens with plants in the rich soil of our gardens; and in these flatter stamens the oxidized juices assumed perhaps a livelier yellow than even the central stamens themselves. If the flowers had fertilized their own ovaries this change would of course have proved disadvantageous, by depriving them entirely of the services of one row of stamens; for the new flattened and petal-like structures lost at once the habit of producing pollen. But their value as attractive organs for alluring the eyes of insects more than counterbalanced this slight apparent disadvantage; and the new petal-bearing blossoms soon outstripped and utterly lived down all their simpler petalless allies. By devoting one outer row of stamens to the function of alluring the fertilizing flies, they have secured the great benefit of perpetual cross-fertilization, and so have got the better of all their less developed competitors. At the same time, the exudations at the base of the petals have assumed the definite form of sweet nectar or honey, a liquid which is mainly composed of sugar, that universal allurer of animal tastes. By this means the plants save their pollen from depredations, and at the same time offer the insects a more effectual, because a more palatable, sort of bribe.

Passing rapidly over these already familiar initial stages, we may go on to those more special and distinctive facts which peculiarly concern the ancestry of the lilies and cereals. It is probable that the nearest modern analogue of the earliest petal-bearing trinary flowers is to be found in the existing alisma tribe, including our own English arrowheads and flowering rushes. As a rule, indeed, it may be said that fresh-water plants and animals tend to preserve for us very ancient types indeed; and all the alismas are marsh or pond flowers of an extremely simple character. They have usually three greenish sepals outside each blossom, inclosing one whorl of three white or pink petals, two or three whorls of three stamens each, and a number of separate ovaries, which are not united, as in the more developed true lilies, into a single capsule, but remain quite distinct, each with its own individual stigma or sensitive surface. Even within this relatively early and simple group, however, several gradations of development may yet be traced. I incline to believe that our English smaller alisma, a not uncommon plant in wet ditches and marshes throughout the whole of Southern Britain, represents the very earliest petal-bearing type in this line of development; indeed, save that its petals are now pinky-white, while those of the original ancestor were almost certainly yellow, we might almost say that the marsh-weed in question was really the earliest petal-hearing plant of which we are in search. It closely resembles in appearance, and in the arrangement of its parts, the buttercups, which are the earliest existing members of the other or quinary division of flowering plants; and in both we seem to get a survival of a still earlier common ancestor, only that in the one the parts are arranged in rows of three, while in the other they are arranged in rows of five; and concomitantly with this distinction go the two or three other distinctions which mark off the two main classes from one another—namely, that the one has leaves with parallel veins, only one seed-leaf to the embryo, and an endogenous stem, while the other has leaves with netted veins, two seed-leaves to the embryo, and an exogenous stem. Nevertheless, in spite of such fundamental differences, we may say that the alismas and the buttercups really stand very close to one another in the order of development. When the two main branches of flowering plants first diverged from one another, the earliest petal-bearing form they produced on one divergent branch was the alisma, or something very like it; the earliest petal-bearing form they produced on the other divergent branch was the buttercup, or something very like it. Hence, whenever we have to deal with the pedigree of either great line, the fixed historical point from which we must needs set out must always be the typical alismas or the typical buttercups. The accompanying diagram will show at once the relation of parts in the simplest trinary flowers, and will serve for comparison at a later stage of our argument with the arrangement of their degraded descendants, the wheats and grasses.

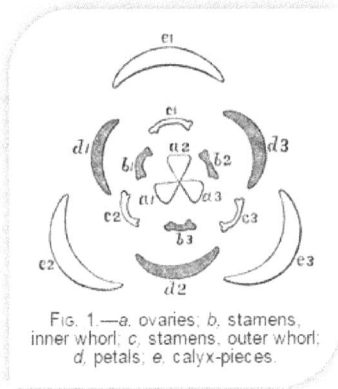

Fig. 1.—a, ovaries; b, stamens, inner whorl; c, stamens, outer whorl; d, petals; e, calyx-pieces.

Our own smaller alisma has a number of ovaries loosely scattered about in its center, as in the buttercups, with two rows of three stamens outside them, and then a single row of three petals, followed by the calyx or inclosing cup of three green pieces. Its close ally the water-plantain, however, shows signs of some advance toward the typical lily form in the arrangement of its ovaries in a single ring, often loosely divisible into three sets. And in the pretty pink flowering rush (not of course a rush at all in the scientific sense) the advance is still more marked in that the number of ovaries is reduced to six, that is to say, two whorls of three each, accompanied by nine stamens, similarly divisible into three rows. In all these very early forms (as in their analogues the buttercups) the main point to notice is this, that there is as yet no regular definiteness in the numerical relations of the parts. They tend to run, it is true, in rows of three; but often these rows are so numerous and so confused that nature loses count, so to speak, and it is only in their higher and more developed members that we begin to arrive at any distinct symmetry, such as that of the flowering rush. Even here, the symmetry is far from being so perfect as in the later lilies. There are, however, a few very special members of the alisma family in which the approach to the true lilies is even greater. These are well represented in England by our own common arrowgrasses—inconspicuous little green flowers, with three calyx-pieces, three petals, six stamens, and either six or three ovaries. Here, too, the ovaries are at first united into a single pistil (as it is technically called), though they afterward separate as they ripen into three or six distinct little capsules. One of our British kinds, the marsh arrowgrass, has almost reached the lily stage of development; for it has three calyx-pieces, three petals, six stamens, and three ovaries, exactly like the true lilies; but it falls short of their full type in the fact that its pistil divides when ripe into separate capsules, whereas the pistil of the lilies always remains united to the very end; and this minute difference suffices, in the eyes of systematic botanists, to make it an alisma rather than a lily. In reality, it ought to be regarded as a benevolent neutral—a surviving intermediate link between the two larger classes.

The specialization which makes the true lilies thus depends upon two points. In the first place, all the parts are regularly symmetrical, except that there are two rows of stamens to each one of the other organs: the common formula being three calyx-pieces, three petals, six stamens, and three ovaries. In the second place, the three ovaries are completely combined together into a single three-celled pistil. The advantage which the lilies thus gain is obvious enough. Then bright petals, usually larger and more attractive than those of the alismas, allure a sufficient number of insects to enable them to dispense with the numerous stamens and ovaries of their primitive ancestors. Moreover, this diminution in number is accompanied by an increase in effectiveness and specialization: for the lilies have only three sensitive surfaces to their pistil, combined on a single stalk; and the honey is usually so placed at its base that the insect cannot fail to brush off pollen at every visit against all three surfaces at once. Again, while the number of ovaries has been lessened, the number of seeds in each has been generally increased, which also marks a step in advance, since it allows many seeds to be impregnated by a single act of pollination. The result of all these improvements, carried further by some lilies than by others, is that the family has absolutely outstripped all others of the trinary class in the race for the possession of the earth, and has now occupied all the most favorable positions in every part of the world. While the alismas and their allies have been so crowded out that they now linger only in a few ponds, marshes, and swamps, to which the more recent lily tribe have not yet had time fully to adapt themselves, the true lilies and their yet more advanced descendants have taken seizin of every climate and every zone upon our planet, and are to be found in every possible position, from the arborescent yuccas and huge agaves of the tropics to the wild hyacinths of our English woodlands and the graceful asphodels of the Mediterranean hill-sides.

The lilies themselves, again, do not all stand on one plane of homogeneous evolution. There are different grades of development still surviving among the class itself. The little yellow gagea which grows sparingly in sandy English fields may be taken as a very fair representative of the simplest and earliest true lily type. It bears a small bunch of little golden flowers, only to be distinguished from the higher alismas by their united ovaries: for though both calyx and petals are here brightly colored, that is also the case in the flowering rushes, and in many others of the alisma group. On the other hand, though it may be said generally of the lilies that their calyx and petals are colored alike— sometimes so much so as to be practically indistinguishable—yet there are many kinds which still retain the greenish calyx-pieces, and that even in the more developed genera. But most of the lilies are far handsomer than gaarea and its allies: even in England itself we have such very conspicuous and attractive flowers as the purple fritillaries, which every Oxford man has gathered by handfuls in the spongy meadows about Iffley lock, with their dark spotted petals converging into a bell, and the nectaries at the base producing each a large drop of luscious honey. Some, like our wild hyacinths, have assumed a tubular shape under stress of insect selection, the better to promote proper fertilization; and at the same time have acquired a blue pigment, to allure the eyes of azure-loving bees. Others have become dappled with spots to act as honey-guides, or have produced brilliant variegated blossoms to attract the attention of great tropical insects. Our British lilies alone comprise such various examples as the lily-of-the-valley, a tubular, white, scented species, adapted for fertilization by moths; the very similar Solomon's-seal; the butcher's-broom; the wild tulip; the star-of-Bethlehem; the various squills; the asparagus; the grape-hyacinth; and the meadowsaffron. Some of them (for example, asparagus and butcher's-broom) have also developed berries in place of dry capsules; and these berries, being eaten by birds which digest the pulp, but not the actual seeds, aid in the dispersion of the seedlings, and so enable the plant to reduce the total number of seeds to three only, or one in each ovary. Among familiar exotics of the same family may be mentioned the hyacinth, tuberose, tulip, asphodel, yucca, and most of the so-called lilies. In short, no tribe supplies us with a greater number of handsome garden flowers, for the most part highly adapted to a very advanced type of insect fertilization.

Properly to understand the development of our existing wheat from this brilliant and ornamental family, as well as to realize the true nature of its relation to allied orders, we must first glance briefly at the upward evolution of the other branches descended from the true lilies, and then recur to the downward evolution which finally resulted in the production of the degenerate grasses. In the main line of progressive development, the lilies gave origin to the amaryllids, familiarly represented in England by the snow-drops and daffodils, a family which is technically described as differing from the lilies in having an inferior instead of a superior ovary — that is to say, with the pistil apparently placed below instead of above the point where the petals and calyx-pieces are inserted. From the evolutionary point of view, however, this difference merely amounts to saying that the amaryllids are tubular lilies, in which the tube has coalesced with the walls of the ovary, so that the petals seem to begin at its summit instead of at its base. The change gives still greater certainty of impregnation, and therefore benefits the race accordingly. At the same time, the amaryllids, being probably a much newer development than the true lilies, have not yet had leisure to gain quite so firm a footing in the world; though on the other hand many of them are far more minutely adapted for special insect fertilization than their earlier allies. They include the so-called Guernsey lilies of our gardens, as well as the huge American aloes which all visitors to the Riviera know so well on the dry hills around Nice and Cannes. The iris family are a similar but rather more advanced tribe, with only three stamens instead of six, their superior organization allowing them readily to dispense with half their complement, and so to attain the perfect trinary symmetry of three sepals, three petals, three stamens, and three ovaries. Among them, the iris and the crocus are circular in shape, but some very advanced types, such as the gladiolus, have acquired a bilateral form, in correlation with special insect visits. From these, the step is not great to the orchids, undoubtedly the highest of all the trinary flowers, with the triple arrangement almost entirely obscured, and with the most extraordinary varieties of adaptation to fertilization by bees or even by humming-birds in the most marvelous fashions. Alike by their inferior ovary, their bilateral shape, their single stamen, their remarkable forms, their brilliant colors, and their occasional mimicry of insect-life, the orchids show themselves to be by far the highest of the trinary flowers, if not, indeed, of the entire vegetable world.

From this brief sketch of the main line of upward evolution from lilies to orchids, we must now return to the grand junction afforded us by the lilies themselves, and travel down the other line of degeneracy and degradation which leads us on to the grasses and the cereals, including at last our own familiar cultivated wheat. Any trinary flower with three calyx-pieces, three petals, six stamens, and a three celled pistil not concealed within an inclosing tube, is said to be a lily, as long as it possesses brightly colored and delicate petals. There are, however, a large number of somewhat specialized lilies with very small and inconspicuous petals, which have been artificially separated by botanists as the rush family, not because they were really different in any important point of structure from the acknowledged lilies, but merely because they had not got such brilliant and handsome blossoms. These despised and neglected plants, however, supply us with the first downward step on the path of degeneracy which leads at last to the grasses, and they may be considered as intermediate stages in the scale of degradation, fortunately preserved for us by exceptional circumstances to the present day. Even among the true lilies, there are some, like the garlic and onion tribe, which show considerable marks of degeneration, owing to some decline from the type of insect fertilization to the undesirable habit of fertilizing themselves. Thus, while our common English rampsons or wild garlic has pretty and conspicuous white blossoms, some other members of the tribe, such as the crow allium, have very small greenish flowers, often reduced to mere shapeless bulbs. Among the true rushes, however, the course of development has been somewhat different. These water-weeds have acquired the habit of trusting for fertilization to the wind, which carries the pollen of one blossom to the sensitive surface of another, perhaps at less trouble and expense to the parent plant than would be necessary for the allurement of bees or flies by all the bribes of brilliant petals and honeyed secretions. To effect this object, their stamens hang out pensile to the breeze, on long, slender filaments, so lightly poised that the merest breath of air amply suffices to dislodge the pollen: while the sensitive surface of the ovaries is prolonged into a branched and feathery process, seen under the microscope to be studded with adhesive glandular knobs, which readily catch and retain every golden grain of the fertilizing powder which may chance to be wafted toward them on the wings of the wind. Under such circumstances, the rush kind could only lose by possessing brightly colored and attractive petals, which would induce insects uselessly to plunder their precious stores: and so all those rushes which showed any tendency in that direction would soon be weeded out by natural selection; while those which produced only dry and inconspicuous petals would become the parents of future generations, and would hand on their own peculiarities to their descendants after them. Thus the existing: rushes are all plain little lilies with dry, brownish flowers, specially adapted to wind-fertilization alone.

Among the rushes themselves, again, there are various levels of retrogressive development—retrogressive, that is to say, if we regard the lily family as an absolute standard: for the various alterations undergone by the different flowers are themselves adaptive to their new condition, though that condition is itself decidedly lower than the one from which they started. The common rush and its immediate congeners resemble the lilies from which they spring in having several seeds in each of the three cells which compose their pistil. But there is an interesting group of small grass-like plants, known as wood-rushes, which combine all the technical characteristics of the true rushes with a general character extremely like that of the grasses. They have long, thin, grass-like blades in the place of leaves; and, what is still more important, as indicating an approach to the essentially one-seeded grass tribe, they have only three seeds in the flower, one to each cell of the capsule. These seeds are comparatively large, and are richly stored with food-stuffs for the supply of the young plantlet. One such richly supplied embryo is worth many little unsupported grains, since it stands a much better chance than they do of surviving in the struggle for existence. The wood-rushes may thus be regarded as some of the earliest plants among the great trinary class to adopt those tactics of storing gluten, starch, and other food-stuffs along with the embryo, which have given the cereals their acknowledged superiority as producers of human food. They are closely connected with the rushes, on the one hand, by sundry intermediate species which possess thin leaves instead of cylindrical, pithy blades; and they lead on to the grasses, on the other, by reason of their very grass-like foliage, and their reduced number of large, well-furnished, starchy seeds.

In another particular, the rush family supplies us with a useful hint in tracing out the pedigree of the grasses and cereals. Their flowers are, for the most part, crowded together in large tufts or heads, each containing a considerable number of minute separate blossoms. Even among the true lilies we find some cases of such crowding in the hyacinths and the squills, or, still better, in the onion and garlic tribe. But, with the wind-fertilized rushes, the grouping together of the flowers has important advantages, because it enables the pollen more easily to fix upon one or other of the sensitive surfaces, as the stalks sway backward and forward before a gentle breeze. Among yet more developed or degraded wind-fertilized plants, this crowding of the blossoms becomes even more conspicuous. A common American rush-like water-plant, known as eriocaulon, helps us to bridge over the gap between the' rushes and such compound flowers as the sedges and grasses. Eriocaulon and its allies have always one seed only in each cell of the pistil; and they have also generally a very delicate corolla and calyx, of from four to six pieces, representing the original three sepals and three petals of the lilies and rushes. But their minute blossoms are closely crowded together in globular heads, the stamens and pistils being here divided in separate flowers, though both kinds of flowers are combined in each head. From an ancestral form not unlike this, but still more like the wood-rushes, we must get both our sedges and our grasses. And though the sedges themselves do not stand in the direct line of descent to wheat and the other cereals, they are yet so valuable as an illustration from their points of analogy and of difference that we must turn aside for a moment to examine the gradual course of their evolution.

The simplest and most primitive sedges now surviving, though very degenerate in type, yet retain some distinct traces of their derivation from earlier rush-like and lily-like ancestors. In the earliest existing type, known as scirpus, the calyx and petals, which were brightly colored in the lilies, and which were reduced to six brown scales in the rushes, have undergone a further degradation to the form of six small, dry bristles, which now merely remain as rudimentary relics of a once useful and beautiful structure. In some species of scirpus, too, the number of these bristles is reduced from six to four or three. There is still one whorl of three stamens, however; but the second whorl has disappeared; while the pistil now contains only one seed instead of three; though it still retains some trace of the original three cells in the fact that there are three sensitive surfaces, united together at their base into one stalk or style. Each such diminution in the number of seeds is always accompanied by an increase in the effectiveness of those which remain; the difference is just analogous to that between the myriad ill-provided eggs of the cod, whose young fry are for the most part snapped up as soon as hatched, and the two or three eggs of birds, which watch their brood with such tender care, or the single young of cows, horses, and elephants, which guard their calves or foals almost up to the age of full maturity. What the bird or the animal effects by constant feeding with worms or milk, the plant effects by storing its seed with assorted food-stuffs for the sprouting embryo.

In the more advanced or more degenerate sedges we get still further differentiation for the special function of wind-fertilization. Take as an example of these most developed types, on this line of development, the common English group of carices. In these the flowers have absolutely lost all trace of a perianth (that is to say, of the calyx and petals), for they do not possess even the six diminutive bristles which form the last relics of those organs in their allies, the scirpus group. Each flower is either male or female—that is to say, it consists of stamens or ovaries alone. The male flowers are represented by a single scale or bract, inclosing three stamens; and in some species even the stamens are reduced to a pair, so that all trace of the original trinary arrangement is absolutely lost. The female flowers are represented by a single ovary, inclosed in a sort of loose bag, which may perhaps be the final rudiment of a tubular, bell-shaped corolla like that of the hyacinth. This ovary contains a single seed, but its shape is often triangular, and it has usually three stigmas or sensitive surfaces, thus dimly pointing back to the three distinct cells of its lily-like ancestors, and the three separate ovaries of its still earlier alisma-like progenitors. In many species, however, even this last souvenir of the trinary type has been utterly obliterated, the ovary having only two stigmas, and assuming a flattened, two-sided shape. In all the carices the flowers are loosely arranged in compact spikes and spikelets, with their mobile stamens hanging out freely to the breeze, and their feathery stigmas prepared to catch the slightest grain of pollen which may happen to be wafted their way by any passing breath of air. The varieties in their arrangement, however, are almost as infinite among the different species as those of the grasses themselves; sometimes the male and female flowers are produced on separate plants; sometimes they grow in separate spikes on the same plant; sometimes the same spike has male flowers at the top and female at the bottom; sometimes the various flowers are mixed up with one another at top and bottom, a regular hotch-potch of higgledy-piggledy confusion. But all the sedges alike are very grass-like in their aspect, with thin blades by way of leaves, and blossoms on tall heads, as in the grasses. In fact, the two families are never accurately distinguished by any except technical botanists; to the ordinary observer, they are all grasses together, without petty distinctions of genus and species. Like the grasses, too, the sedges are mostly plants of the open, wind-swept plains or marshy levels, where the facilities for wind-fertilization are greatest and most constantly present.[1]

And now, from this illustrative digression, let us hark back again to the junction-point of the rushes, whence alike the sedges and the grasses appear to diverge. In order to understand the nature of the steps by which the cereals have been developed from rush-like ancestors, it will be necessary to look shortly at the actual composition of the flower in grasses, which is the only part of their organism differing appreciably from the ordinary lily type. The blossoms of grasses, in their simplest form, consist of several little green florets, arranged in small clusters, known as spikelets, along a single common axis. Of this arrangement, the head of wheat itself offers a familiar and excellent example. If we pull to pieces one of the spikelets composing such a head, we find it to consist of four or five distinct florets. Omitting special features and unnecessary details, we may say that each floret is made up of two chaffy scales, known as pales, and representing the calyx, together with a pair of small white petals known as lodicules, three stamens, and an ovary with two feathery styles. Moreover, the two pales or calyx-pieces are not similar and symmetrical, for the outer one is simple and convex, while the inner one is apparently double, being made up of two pieces rolled into one, and still possessing two green midribs, which show distinctly like ribs on its flat outer surface. Here, it will immediately be apparent, the traces of the original trinary arrangement are very slight indeed.

But when we come to inquire into the rationale and genesis of these curiously one sided flowers, it is not difficult to see that they have been ultimately derived from trinary blossoms of the rush-like type. The first and most marked divergence from that type, for which the analogy of the sedges has already prepared us, is the reduction of the ovary to a single one-seeded cell, whose ripe, fruity form is known as a grain. At one time, we may feel pretty sure, there must have existed a group of nascent grasses, which only differed from the wood-rush genus in having a single-celled ovary instead of a three celled pistil with one seed in each cell; and even the ovary of this primitive grass must have retained one mark of its trinary origin in its possession of three styles to its one grain, thus pointing back (as most sedges still do) to its earlier rush-like origin. That hypothetical form must have had three sepals, three petals, six stamens, and one three-styled ovary. But the peculiar shape of modern grass-flowers is clearly due to their very spiky arrangement along the edge of the axis. In the wood-rushes and the sedges, we see some approach to this condition; but in the grasses, the crowding is far more marked, and the one-sidedness has accordingly become far more conspicuous. Suppose we begin to crowd a number of wind-fertilized lily-like flowers along an axis in this manner, taking care that the stamens and the sensitive feathery styles are always turned outward to catch the breeze (for otherwise they will die out at once), what sort of result shall we finally get?

In the first place, the calyx, consisting of three pieces, will stand toward the crowded stem or axis in such a fashion that one piece will be free and exterior, while two pieces will be interior and next the stem, thus:

O
a a
a

Now, the effect of constant crushing in this direction will be that the two inner calyx-pieces will be slowly dwarfed, and will tend to coalesce with one another; and this is what has actually happened with the inner pale of wheat and of other grasses, though the midribs of the two originally separate pieces still show on the compound pale, like dark-green lines down its center. Thus, in the fully developed grasses, in place of a trinary calyx, we get two chaffy scales or pales, the outer one representing a single sepal, and the inner one, which has been dwarfed by pressure against the stem, representing two sepals rolled into one, with two midribs still remaining as evidence of their original distinctness.

Next, in the case of the petals, which alternate with the sepals of the calyx, the relation to the stem is exactly reversed; for we have here two petals free and exterior, with one interior petal crowded closely against the axis, thus:

O

a

a a

Here, then, the two external petals will be saved, exactly as the one external sepal was saved in the case of the calyx; and these two petals are represented by the very small white lodicules under the outer pale in our existing wheats and grasses. On the other hand, the inner petal, jammed in between the grain and the inner pale (with the stem at its back), has been utterly crushed out of existence, partly because of its very small size, partly because of its functional uselessness, and partly because it had no other part with which to coalesce, and so to save itself as the inner sepals had managed to do. Moreover, it must be remembered that the sepals do still perform a useful service in protecting the young flower before it opens, and in keeping out noxious insects during the kerning or swelling of the grain; whereas the lodicules or rudimentary petals are now apparently quite functionless; and so we may congratulate ourselves that they are there at all, to preserve for us the true ground-plan of the floral architecture in grasses. Indeed, they have not survived by any means in all grasses; among the smaller and more degraded kinds they are often wholly wanting, having been quite crushed out between the calyx and the grain. It is only the larger and more primitive types that still exhibit them in any great perfection. On the other hand, one group of very large exotic grasses, the bamboos, has three regular petals, thus clearly showing the descent of the family as a whole from rush-like ancestors, and also obviously suggesting that the obsolescence of the inner petal in the other grasses is due to their small size and their closely packed minute flowers.

Among the stamens, one-sidedness has not notably established itself, for in wind-fertilized plants they must necessarily hang out freely to the breeze, and therefore they do not get much crowded between the other parts. A few grasses still even retain their double row of stamens, having six to each floret; but most of them have only one whorl of three. In some of the lower and more degraded forms, however, even the stamens have lost their trinary order, and only two now survive. This is the case in our own very degenerate little sweet vernal-grass, the plant which imparts its delicious fragrance to new mown hay. But in the cereals and in most other large species the three stamens still remain in undiminished effectiveness to the present day.

Finally, we come to the most important part of all, the ovary. This part, alternating with the stamens, has the same arrangement of styles relatively to the axis as in the case of the petals; and it has undergone precisely the same sort of abortive distortion. The two outer styles, hanging freely out of the calyx, have been preserved like the two outer lodicules; but the inner one, pressed between the grain and the inner pale (with the stem behind it), has been simply crushed out of existence, like its neighbor the inner lodicule.

Thus the final result is that the whole inner portion of the flower (except as regards stamens) has been distorted or rendered abortive by close pressure against the stem (due to the crowding of the florets in the spiky form), while the whole outer portion remains normal and fully developed. We have an outer pale representing a single normal sepal, and an inner pale representing two dwarfed and united sepals; we have two normal outer lodicules or petals, and a blank where the inner petal ought to be; we have three stamens, symmetrically arranged, among the faithless faithful only found; and we have finally two normal outer styles, with a blank in place of the absent inner style.

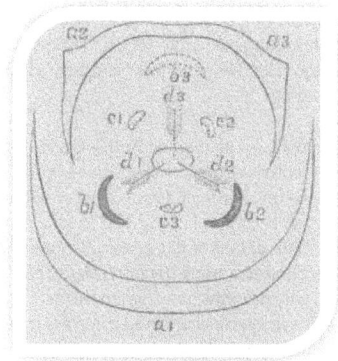

The above diagram, compared with that already given, will make this perfectly clear.

Here, a1 represents the outer pale or normal sepal, while a2 and a3 represent the inner pale composed of the two united sepals. Again, b1 and b2 stand for the two lodicules or surviving petals, while b3 marks the place of the lost petal, now found in the bamboos alone. The stamens are lettered c1, c2, and c3. The two existing styles are shown by d1 and d2, while d3 marks the abortive inner style, now not even present in a rudimentary condition. It will be observed at once that all the outer side is normal, and all the inner side more or less abortive through pressure against the axis.

Thus it will be seen that the line of links which connects the grasses and cereals with the lilies is absolutely unbroken, and that it consists throughout of one continuous course of degradation. At the same time, by this one-sided and spiky arrangement, the grasses secured for themselves an exceptional advantage in the struggle for existence. No other race of small, wind-fertilized plants could compete with them for the possession of the open, wind-swept plains; and over all these they spread far and wide, rapidly differentiating themselves into a vast number of divergent genera and species, each adaptively specialized for some peculiar habitat, soil, or climate. At the present time, the grasses number their kinds by thousands; they extend over the whole world, from the poles to the equator; and they form the general sward or carpet of greenery over by far the larger portion of the terrestrial globe. Even in Britain alone, with our poor little insular flora, a mere fragment of that belonging to the petty European Continent, we number no less than forty-two genera of grasses, distributed into more than one hundred species. In fact, what may fairly be called degradation from one point of view may fairly be called adaptation from another. The organization of the grasses is certainly lower than that of the lilies, but it fits them better for that station of life to which it has pleased Nature to assign them.

The various kinds of grasses differ very little from one another in general plan; the flower in almost all is constructed strictly on the lines above mentioned; and the leaves in almost all are just the same soft, pensile blades, making them into the proper greensward for open, unwooded, wind-swept plains. But, like almost all other very dominant families, they have split up into an immense number of kinds, distinguished from one another by minute differences in the arrangement of the florets and the spikelets; and these kinds have again subdivided into more and more minutely different genera and species. One great group, with panicles of a loose character, and very degraded spikelets, has given origin to many southern grasses, from some of which the cultivated millets are derived. Another great group, with usually more spiky inflorescence, has given origin to most of our northern grasses, from some of which the common cereals are derived. This second group has again split up into several others, of which the important one for our present purpose is that of the Hordeineæ, or barley-worts. From one of the numerous genera into which the primitive Hordeineæ have once more split up, our cultivated barleys take their rise; from another, which here demands further attention, we get our cultivated wheats.

The nearest form to true wheat now found wild in the British Isles is the creeping couch-grass, a perennial closely agreeing in all essential particulars of structure with our cultivated annual wheats. But in the South European region we find in abundance a large series of common wild annual grasses, forming the genus Ægilops of technical botany, and exactly resembling true wheat in every point except the size of the grain. One species of this genus, Ægilops ovata, a small, hard, wiry annual, is now pretty generally recognized among botanists as the parent of our cultivated corn. There was a good reason, indeed, why primitive man, when he first began to select and rudely till a few seeds for his own use, should have specially affected the grass tribe. No other family of plants has seeds richer in starches and glutens, as indeed might naturally be expected from the extreme diminution in the number of seeds to each flower. On the other hand, the flowers on each plant are peculiarly numerous; so that we get the combined advantages of many seeds, and rich seeds, so seldom to be found elsewhere, except among the pulse family. The experiments conducted by the Agricultural Society in their College Garden at Cirencester have also shown that careful selection will produce large and rich seeds from Ægilops ovata, considerably resembling true wheat, after only a few years' cultivation.

Primitive man, of course, did not proceed nearly so fast as that. Of the very earliest attempts at cultivation of Ægilops, all traces are now lost, but we can gather that its tillage must have continued in some unknown Western Asiatic region for some time before the neolithic period; for in that period we find a rude early form of wheat already considerably developed among the scanty relics of the Swiss lake-dwellings. The other cultivated plants by which it is there accompanied and the nature of the garden-weeds which had followed in its wake point back to Central or Western Asia as the land in which its tillage had first begun. From that region the Swiss lake-dwellers brought it with them to their new home among the Alpine valleys. It differed much already from the wild Ægilops in size and stature; but at the same time it was far from having attained the stately dimensions of our modern corn. The ears found in the lake-dwellings are shorter and narrower than our own; the spikelets stand out more horizontally, and the grains are hardly more than half the size of their modern descendants. The same thing is true in analogous ways with all the cultivated fruits or seeds of the stone age; they are invariably much smaller and poorer than their representatives in existing fields or gardens. From that time to this the process of selection and amelioration has been constant and unbroken, until in our own day the descendants of these little degraded lilies, readapted to new functions under a fresh régime, have come to cover almost all the cultivable plains in all civilized countries, and supply by far the largest part of man's food in Europe, Asia, America, and Australia.

1 - The sedges are not, in all probability, a real natural family, but are a group of heterogeneous, degraded lilies, containing almost all those kinds in which the reduced florets are covered by a single conspicuous glume-like bract. It will be seen from the sequel that these bracts are not truly analogous to the glumes or outer palese of grasses.

FROM BUTTERCUPS TO MONK'S-HOOD

To look at these queer, irregular blue flowers, growing on a long and handsome spike in the old-fashioned garden border, nobody would ever dream of saying that they were in reality altered and modified buttercups. And yet that is just what they really are, with all the marks of their curious pedigree still clearly impressed upon their very form. Pull one of the blue blossoms off, and pick it carefully to pieces, and you will see how strangely and profoundly it has been distorted by insect selection. Monk's-hood is most essentially a bee-flower, and in examining it we see the results of bee action plainly set forth in every organ. If we pick a common meadow buttercup for comparison with it, we shall be able to see exactly wherein the two flowers differ, as well as why the one has gained an advantage in the struggle for existence over the other.

The outside whorl of the buttercup consists, of course, of five separate greenish sepals, which together make up its calyx. Inside the sepals come the five golden petals composing the cup-shaped corolla; and inside the petals, again, come the numerous stamens, and the equally numerous carpels or unripe fruits, each containing a single solitary little seed. Moreover, all these parts are regularly and symmetrically arranged round a common center, so as to form a series of concentric whorls. But when we look at the monk's-hood we see no such simple and orderly arrangement in its architectural plan. At first sight, we recognize no distinct sepals or petals: and the colored organs that take their place are very irregular in shape, and disposed in an unsymmetrical fashion—or rather, to speak more correctly, their symmetry is not radial, but bilateral. When we begin to pull our blue blossom to pieces, however, we gradually recognize the various parts of which it is composed. First of all come five sepals, not greenish as in the buttercup, but bright blue; and not all alike, but specially modified to fulfill their separate functions. The uppermost sepal of all is helmet-shaped, and it forms the curious cowl which gained the plant its suggestive name from our mediæval ancestors. The two side sepals, to right and left, are flatter and straighter, but very broad, while the two lowest of all are comparatively small and narrow. The whole five are bright blue in color. Pull off these petal-like sepals, and you come to the real petals beneath them. At first you can hardly find them at all; you see only two long blue horns, covered till now by the helmet-shaped upper sepal or cowl, and each with a queer cup-like sac at its extremity, containing a small drop of clear fluid. That fluid is honey, but I should advise you to be careful in tasting it not to bite off any of the flower, for monk's-hood is the plant from which we get the now famous poison, aconitine; and a very little of it goes a long way. Unlike as they are to the familiar yellow petals of the buttercup, one can still gather from their position that the two long horns are really petals. But where are the three others? Well, you must look rather close to find them, and perhaps even then you won't succeed after all; for sometimes the three lower petals have disappeared altogether, being suppressed by the plant as of no further use to it. In this particular specimen, however, they still survive as mere relics or rudiments, three little narrow blue blades, not nearly as big as a gnat's wing, placed alternately to the lower sepals. As for the stamens, they are still present about as numerously as in the buttercup; whereas the carpels, or fruit-pieces, are reduced to three only, which in the ripe seed-vessels here on the lower and older part of the spike grow into long pods or follicles, each containing several seeds.

Thus, then, the flower of monk's-hood agrees fundamentally with the flower of the buttercup; while, at the same time, it has undergone some very singular and suggestive modifications. In both there are five sepals; but in the buttercup all five are alike, and all five are greenish; whereas in the monk's-hood they have acquired different shapes, exactly fitting them to the bee's body, and they have become blue, because blue is the favorite color of bees. Again, in both there are five petals; but in the buttercup all five are similar and yellow, and all five secrete a drop of honey at the base; whereas in the monk's-hood two of them have become long and narrow specialized nectaries, while the other three, being no longer needed, have grown obsolete or nearly so. Once more, the stamens remain the same; but the carpels have been immensely reduced in number, at the same time that the complement of seeds in each has been greatly increased by way of compensation.

Well, how are we to account for these peculiar modifications ? Entirely by the action of the fertilizing bees. The secret of the monk's-hood depends, in the first place, upon the fact that its flowers are clustered into a spike, instead of growing in solitary isolation at the end of the stem, as in the common buttercups. Now, Mr. Herbert Spencer has pointed out that solitary terminal flowers are always radially symmetrical, and never one-sided, because the conditions are the same all round, and the visiting insects can light upon them equally from every side. But flowers which grow sideways from a spike are very apt to become bilaterally symmetrical ; indeed, whenever they are not so, one can always give an easy explanation of their deviation from the rule. Probably the blossoms of the monk's-hood began by arranging themselves in a long and handsome spike, so as more readily to attract the eyes of insects; and that was the real starting-point of all their subsequent modifications. Or, to put the same thing more literally, those monk's-hoods which happened to grow spike-wise succeeded best in attracting the bees, and therefore were most often fertilized in the proper manner. Next, we may suppose, the large green sepals, being much exposed to view, began to acquire a bluish tinge, as all the upper parts of highly developed plants are apt to do ; and the bluer they became, the more conspicuous they looked, and therefore the better they got on in competition with their neighbors, especially since bees are particularly fond of blue. As each bee would necessarily light on the middle or lower portion of the flower, he would begin by extracting the honey from the two upper petals; but it would be rather awkward for him to turn round head downward, and suck the nectaries of the three bottom ones. Hence, in course of time, especially after the flower began to acquire its present shape, the two top petals became specialized as nectaries, while the three lower ones gradually atrophied, since the colored sepals had practically usurped their attractive function. But as the flower can only succeed by being fertilized, all these changes must have been really subordinate to the great change which was simultaneously going on in the mechanism for insuring fertilization. Slowly the blossoms altered to the bilateral shape—they adapted themselves by the bee's unconscious selection to the insect's form. The uppermost sepal grew into the hood, so arranged that the bee must get under it in order to reach the long nectaries containing their copious store of honey. At the same time the bee must brush against the stamens, and cover his breast with a stock of adhesive pollen-grains. When he flies away to the next flower he carries the pollen with him and, as he rifles the nectaries in the second blossom, he both deposits pollen from the last plant upon the sensitive surface of the carpels in this, and also collects a fresh lot of pollen to fertilize whatever other flower he may next favor with a call. The increased certainty of fertilization thus obtained enables the plant to dispense with some of the extra carpels which its buttercup ancestors once possessed; and, by lessening the number to three, it manages to get the whole set impregnated at a single visit. But, as three seeds would be a small number to depend upon in a world of overstocked markets and adverse chances, it makes up for the diminution of its carpels by largely increasing the stock of seeds in each.

Thus the whole shape and arrangement of the monk's-hood bear distinct reference to the habits and tastes of the fertilizing bees. It is a mountain plant by origin, belonging to a tribe which took its rise among the great central chains of Europe and Asia, and these Alpine races are usually highly developed in adaptation to insect fertilization, because they depend more absolutely upon a few upland species than do the eclectic flowers of the plains, which may be impregnated haphazard by a dozen different flies, or moths, or beetles. We can still dimly trace many of the links which connect it with very simple and primitive buttercups, if not directly, at least by the analogy of other plants. For all the buttercup tribe show us regular gradations in the same direction. The simplest kinds are round, yellow, and many-carpeled, like the buttercups. Then those species which display their sepals largely have dwarfed petals, like hellebore and globe-flower, or have lost them altogether, like marsh-marigold, which trusts entirely for color display to its big golden calyx. The still higher anemones have the sepals white, red, or blue; and the very advanced columbine has all the petals spurred, and developed into nectaries, like those of monk's-hood. But columbine still keeps to single terminal flowers, so that here the five petals remain regular and circularly symmetrical, though the carpels are reduced to five. Fancy a number of such columbine-flowers crowded together on a spike, however, and you can readily picture to yourself by rough analogy the origin of monk's-hood. The sepals would now become the most conspicuous part; the two upper petals would alone be useful in insuring fertilization, and the lower ones would soon shrivel away from pure disuse. The development of the hood and the lengthening of the upper petals would easily follow by insect selection. It is a significant fact that our only other spiked buttercup, the larkspur, has equally irregular and bilateral flowers, though its honey is concealed in a long spur formed by the petals, and accessible to but one English insect, the humble-bee.

Grant Allen – A Short Biography

Charles Grant Blairfindie Allen was born on February 24th, 1848 at Alwington, near Kingston, Canada West (now part of Ontario). He was the second son of the Rev. Joseph Antisell Allen, a Protestant minister from Dublin, Ireland and Catharine Ann Grant, the daughter of the fifth Baron of Longueuil.

Grant was educated at home until he was thirteen at which time the family moved, initially to the United States, then France and finally settling in the United Kingdom.

Whilst growing up the family background was obviously religious but Grant developed his own views on life and the world and turned to agnosticism and socialism.

He was educated at King Edward's School in Birmingham and Merton College in Oxford. After graduating, Grant studied in France and also taught at Brighton College. By 1870, still only in his mid-twenties, he became a professor at Queen's College, a black college in Jamaica.

Whilst in Jamaica Grant met and married his first wife Ellen Jerrard in 1873 and they produced a son five years later; Jerrard Grant Allen, who grew up to become a theatrical agent/manager.

In 1876 Grant and his family left Jamaica to return to England with both the talent and ambition to become a writer.

He quickly turned to writing essays, gaining a reputation for his work on science and literary works. An early article, 'Note-Deafness' a description of what is now called amusia, was published in 1878 in the learned journal Mind and was cited approvingly by Oliver Sacks very recently.

From essays in magazines and journals he now turned to books, initially on scientific subjects. These include Physiological Æsthetics 1877 and Flowers and Their Pedigrees 1886.

His first major influence was associationist psychology, as then expounded by Alexander Bain and Herbert Spencer, the latter is often considered the most important individual in the transition from associationist psychology to Darwinian functionalism. In Grant's many articles on flowers and perception in insects, Darwinian arguments now replaced the old Spencerian terms.

On a personal level, a long friendship that started when Grant met Herbert Spencer on his return from Jamaica, turned eventually to one of unease over its long course. Grant was to write a critical and revealing biographical article on Spencer that was published after Spencer was dead.

In the early 1880's Grant began to assist Sir W. W. Hunter in his Gazeteer of India. It is at this time that Grant now turned his full attention away from the factual and towards the world of imagination and fiction.

Between this shift to fiction in 1884 and his death fifteen years later Grant was to write about 30 novels.

Many were adventure novels which were very common in the late Victorian period as writers turned their literary talents to the voracious appetites of the weekly or monthly serial magazines.

Some however were to cause quite a stir. For instance in 1895 Grant took the subject of children born out of wedlock as his subject matter. The result was The Woman Who Did, that suggested, indeed pushed, for its time, certain quite startling views on marriage and related areas. In keeping with his then glowing reputation it became a bestseller despite it being seemingly at odds with society's unease at its provocative subject matter.

Interestingly Grant wrote novels under female pseudonyms. One of these was the short novel The Type-writer Girl, which he wrote under the name Olive Pratt Rayner.

Another work, The Evolution of the Idea of God 1897, propounding a theory of religion on heterodox lines, has the disadvantage of endeavoring to explain everything by one theory. This "ghost theory" was often seen as a derivative of Herbert Spencer's theory. However, at the time, it was well known and brief references to it can be found in a review by Marcel Mauss, Durkheim's nephew, in the articles of William James and in the works of Sigmund Freud. The young G. K. Chesterton wrote on what he considered the flawed premise of the idea, arguing that the idea of God preceded human mythologies, rather than developing from them. Chesterton said of Grant Allen's book on the evolution of the idea of God "it would be much more interesting if God wrote a book on the evolution of the idea of Grant Allen".

From this and other instances, it can be seen that his work was in debate and whether agreed with or not could always ensure a lively discussion.

Grant also helped to pioneer science fiction, with the 1895 novel The British Barbarians. This book, was published at about the same time as H. G. Wells was to publish The Time Machine. The plots are quite different but both describe time travel. A few years later his short story The Thames Valley Catastrophe (published 1901 in The Strand magazine) describes the destruction of London by a massive volcanic eruption. Whilst the premise now may seem outlandish, at the time genuine panic and concern set in as, like his contemporary, Jules Verne, much of great science fiction writing is rooted in a plausibility that is set out very convincingly.

In detective fiction too his works include female detectives, very much an innovation in the young genre and his gentleman rogue, Colonel Clay, is seen as a forerunner to other, perhaps more famous characters, by other later writers.

In 1881 he had settled at Dorking, where he took great delight in botanical walks in the woods and sandy heaths. He never enjoyed particularly good health and so almost every winter he would depart for milder climes, to winter in the south of Europe, usually at Antibes, though occasionally as far as Algiers and Egypt.

In 1892 he bought land almost on the summit of Hind Head, and built himself a charming cottage which he called the Croft. Here he found that it was possible to endure the vagaries of the English winter and in landscape more beautiful and wilder than at Dorking and that his long scientific training could better appreciate.

His growing re-discovery and interest in art in the later part of his life allowed him to blend together literature, art and history in a series of guide books on Paris, Florence, Venice, and the cities of Belgium.

On October 25th 1899 Grant Allen died at his home in Hindhead, Haslemere, Surrey, England. He died just before finishing Hilda Wade. The novel's final episode, which he dictated to his friend, doctor and neighbour Sir Arthur Conan Doyle from his bed appeared under the appropriate title, The Episode of the Dead Man Who Spoke in the Strand Magazine in 1900.

Grant Allen is rarely heard of today, although an occasional short story can be heard on the radio or reprinted among magazine enthusiasts but in his time he did much to entertain the masses and push several genres along a richer journey they are still proceeding on today.

Grant Allen – A Concise Bibliography

Physiological Æsthetics. 1877
The Colour-Sense: Its Origin and Development. 1879
Evolutionist at Large. 1881
Vignettes from Nature. 1881
The Colours of Flowers. 1882
Colin Clout's Calendar. 1883
Flowers and Their Pedigrees. 1883
Philistia. 1884
Strange Stories. Short Stories. 1884
Babylon. A novel in 3 volumes. 1885
For Mamie's Sake. 1886
In All Shades. 1886
The Beckoning Hand & Other Stories. 1887
This Mortal Coil: A Novel. 1888
Force and Energy. 1888
The Devil's Die. 1888
The White Man's Foot. 1888
Falling in Love. 1889
The Tents of Shem. 1889
Wednesday the Tenth. 1890
The Great Taboo. 1890
Dumaresq's Daughter. 1891
What's Bred in the Bone. 1891
The Duchess of Powysland. 1892

The Scallywag. 1893
Michael's Crag. 1893
The Lower Slopes. 1894
Post-Prandial Philosophy. 1894
The British Barbarians. 1895
At Market Value. 1895
The Story of the Plants. 1895
The Desire of the Eyes. 1895
The Woman Who Did. 1895
The Jaws of Death. 1896
A Bride from the Desert. 1896
Under Sealed Orders. 1896
Moorland Idylls. 1896
An African Millionaire. Colonel Clay's novel. 1897
The Evolution of the Idea of God. 1897
Paris. 1897
The Type-writer Girl. (as Olive Pratt Rayner) 1897
Tom, Unlimited. (as Martin Leach Warborough) 1897
Flashlights on Nature. 1898
The Incidental Bishop. 1898
Venice. 1898
The European Tour. 1899
A Splendid Sin. 1899
Miss Cayley's Adventures. 1899
Twelve Tales: With a Headpiece, a Tailpiece, and an Intermezzo. 1899
Hilda Wade (finished by Arthur Conan Doyle). 1900
Linnet. 1900
The Backslider. 1901
Sir Theodore's Guest & Other Stories. 1902
Evolution in Italian Art. 1908
The Hand of God. 1909
The Plants. 1909

Short Stories
The Empress of Andorra. 1878
My New Year Among the Mummies. 1878
Lucretia. 1879
My Circular Tour. 1880
A Ballade of Evolution. 1880
Ram Das of Cawnpore. 1880
The Chinese Play at the Haymarket. 1880
The Senior Proctor's Wooing. 1881
Pausodyne. 1881
Caribbean Twelve Per Cents. 1882
An Episode in High Life. 1882
Mr Chung. 1882
Isadine and I. 1883
The Backsider. 1883
The Reverend John Creedy. 1883
The Foundering of the Fortuna. 1883

The Third Time
The Gold Wulfric
My Uncle's Will. 1884
Carvalho. 1884
The Mysterious Occurence in Piccadilly. 1884
Dr Greatex's Engagement. 1884
Hugh Portledown's Return from Normandy. 1884
The Child of Phalanstery. 1884
The Curate of Churnside. 1884
John Cann's Treasure. 1884
Olga Davidoff's Husband. 1884
The Search Party's Find. 1885
The Two Carnegie's. 1885
Professor Milliter's Dilemma. 1885
In Strict Confidence. 1885
The Beckoning Hand. 1885
The Third Time. 1886
Harry's Inheritance. 1886
The Gold Wulfric. 1886
Mr Pierpoint's Repentance. 1886
Claude Tyack's Ordeal. 1887
Leonard's Recovery. 1887
A Social Difficulty. 1887
Dr Palliser's Patient. 1888
My Christmas Eve at Marzin. 1888
The Sultan's Sister. 1888
His First Crime. 1889
The Mayfield Mystery. 1889
Andre Canivet's Curse. 1890
Old Margaret. 1890
My One Gorilla. 1890
Dick Prothero's Luck. 1890
A Deadly Dilemna. 1891
Jerry Stokes. 1891
Selwyn Utterton's Nemesis. 1891
General Passavant's Will. 1891
The Briefless Barrister. 1891
Melissa's Tour. 1891
Karen – A Canadian Romance. 1891
The Prisoner of Assiout. 1891
The Abbe's Repentance. 1891
Masie Bowman's Fate. 1891
Naomi's Christmas Eves. 1891
That Friend of Sylvia's. 1892
The Conscientious Burglar. 1892
The Minor Poet. 1892
The Governor's Story. 1892
The Pot Boiler. 1892
The Great Ruby. 1892
Ewen Murray's Swim. 1892
Ivan Greet's Masterpiece. 1892

Pallinghurst Barrow. 1892
Langalula. 1893
The Assasin's Knife. 1893
The Artist and the Penny-a-Liner. 1893
A Casual Conversation. 1893
How To Succeed in Literature. 1893
Torrigiano. 1893
A Modern Sibyl: A Florentine Sketch. 1893
Nemesis Wins. 1894
Cecca's Lover. 1894
A Self Respecting Servant. 1894
Passiflora Sanguinea. 1894
An Excellent Match. 1894
Major Kinfaun's Marriage. 1894
Grateful Joe. 1894
An Idyll of the Ice. 1894
Criss Cross Love. 1894
Poor Little Soul. 1894
Amour de Voyage. 1894
The Dynamiter's Sweetheart. 1894
A Triumph of Civilisation. 1894
Dr Wardroper's Lie. 1894
The Miraclous Explorer. 1894
Leon and Leonie. 1895
A Comic Emotion. 1895
Joe's Rascality. 1895
Evelyn Moore's Poet. 1895
Frasine's First Communion. 1895
TheDead Man Speaks. 1895
A Study From the Nude. 1895
Cecca's Choice. 1895
The Desire of the Eyes. 1895
The Making of a Poet. 1895
The Man From Cumbrae. 1895
Fogo Skerries. 1895
The Great Californian Heiress. 1895
Cap'n Tom Woolley. 1895
The Girl at the Fair. 1895
Love's Old Dream. 1895
A Modern Pygmalion. 1895
A Bride From the Desert. 1895
The Practical Test. 1896
A Confidential Communication. 1896
The Great Temperance Preacher. 1896
A Day on the River. 1896
The Episode of the Mexican Seer. 1896
A Midsummer Episode. 1896
The Episode of the Diamond Links. 1896
Omar at Marlow. 1896
A Mere Matter of Standpoint. 1896
Fair Exchange. 1896

The Cowardly Dynamiter. 1896
The Episode of the Old Master. 1896
The Episode of the Tyrolean Castle. 1896
Janet's Nemesis. 1896
Entirely Accidental. 1896
The Episode of the Drawn Game. 1896
Wolverden Tower. 1896
The Episode of the German Professor. 1896
The Episode of the Arrest of the Colonel. 1896
The Episode of the Seldom Gold Mine. 1897
The Camisard's Bride. 1897
The Episode of the Japanned Dispatch Box. 1897
Llanfihangel Skerries. 1897
The Episode of the Game of Poker. 1897
The Episode of the Bertillon Method. 1897
A Lady of Florence. 1897
The Episode of the Old Bailey. 1897
A British Verdict. 1897
A Domestic Tragedy. 1897
A College Charm. 1897
A Freak of Memory. 1897
The Judge's Cross. 1897
The Thames Valley Catastrophe. 1897
The Great Oriental Seer. 1897
The Adventures of the Cantankerous Old Lady. 1898
The Adventure of the Supercilious Attache. 1898
The Pirate of Cliveden Reach. 1898
The Adventure of the Amateur Commission. 1898.
The Adventure of the Impromptu Mountaineer. 1898
Joe's Wife. 1898
The Adventure of the Urbane Old Gentleman. 1898
The Adventure of the Unobtrusive Oasis. 1898
The Adventure of the Pea Green Patrician. 1898
Isenberg's Regiment. 1898
The Adventure of the Magnificent Maharajah. 1898
A Woman's Hand: A Story. 1898
The Adventure of the Cross Eyed QC. 1898
The Christmas Eve Concert. 1898
The Adventure of the Oriental Attendant. 1899
Joseph's Dream. 1899
The Adventure of the Unprofessional Detective. 1899
Hobbling Mary. 1899
The Episode of the Patient Who Disappointed Her Doctor. 1899
The Episode of the Gentleman Who Had Failed For Everything. 1899
The Episode of the Wife Who Did Her Duty. 1899
The Episode of the Man Who Would Not Commit Suicide. 1899
A Regrettable Error. 1899
Peace-At-Any-Price Bill. 1899
The Episode of the Letter with a Basingstoke Post-Mark. 1899
The Episode of the Stone That Looked About It. 1899
His Ways Inscrutable. 1899

The Episode of the European With A Kaffir Heart. 1899
The Episode of the Lady Who Was Very Exclusive. 1899
The Episode of the Guide Who Knew the Country. 1899
Luigi and the Salvationist. 1899.
A Christmas Adventure. 1899.
The Episode of the Officer Who Understood Perfectly. 1900
Meriel Stanley, Poacher. 1900
The Episode of the Dead Man Who Spoke. 1900
A Question of Colour. 1900
Fra Benedett's Medal: A Story. 1900
The Temple of Fate: A Fable. 1900
The Way to Keronan. 1902
Lucy Lockett. 1902
Spencerian. 1904

Articles

1878. Hellas and Civilization, Gentleman's Magazine, Vol. CCXLIII
1878. Nation-making: A Theory of National Characters, Gentleman's Magazine, Vol. CCXLIII
1878. The Origin of Fruits, in Popular Science Monthly Volume 13
1879. Why Do We Eat our Dinner? in Popular Science Monthly Volume 14
1879. A Problem in Human Evolution, in Popular Science Monthly Volume
1879. Pleased with a Feather, in Popular Science Monthly Volume 15
1880. Why Keep India? The Contemporary Review, Vol. XXXVIII
1880. The Growth of Sculpture, The Cornhill Magazine, Vol. XLII
1880. The English Chronicle, Gentleman's Magazine, Vol. CCXLV
1880. The Venerable Bede, Gentleman's Magazine, Vol. CCXLIX
1880. The Dog's Universe, Gentleman's Magazine, Vol. CCXLIX
1880. Evolution and Geological Time, Gentleman's Magazine, Vol. CCXLIX
1880. Geology and History, in Popular Science Monthly Volume 17
1880. Aesthetic Feeling in Birds, in Popular Science Monthly Volume 17
1880. Aesthetic Evolution in Man, in Popular Science Monthly Volume 18
1881. The Story of Wulfgeat, Gentleman's Magazine, Vol. CCLI
1882. An English Shire, Gentleman's Magazine, Vol. CCLII
1882. The Welsh in the West Country, Gentleman's Magazine, Vol. CCLIII
1882. The Colours of Flowers, The Cornhill Magazine, Vol. XLV
1882. An English Weed, The Cornhill Magazine, Vol. XLV
1882. Sir Charles Lyell, in Popular Science Monthly Volume 20
1882. Hyacinth-Bulbs, in Popular Science Monthly Volume 20
1882. Who was Primitive Man? in Popular Science Monthly Volume 22
1883. The Pedigree of Wheat, in Popular Science Monthly Volume 22
1883. From Buttercups to Monk's-Hood, in Popular Science Monthly Volume 23
1883. Honeysuckle, Gentleman's Magazine, Vol. CCLV
1884. The Garden Snail, Gentleman's Magazine, Vol. CCLVI
1884. Our Debt to Insects, Gentleman's Magazine, Vol. CCLVI
1884. Idiosyncrasy, in Popular Science Monthly Volume 24
1884. The Ancestry of Birds, in Popular Science Monthly Volume 24
1884. The Milk in the Cocoa-Nut, in Popular Science Monthly Volume 25
1884. Our Debt to Insects, in Popular Science Monthly Volume 25
1884. Hickory-Nuts and Butternuts, in Popular Science Monthly Volume 25
1884. Queer Flowers, in Popular Science Monthly Volume 26

1885. Food and Feeding, in Popular Science Monthly Volume 26

1885. Concerning Clover, in Popular Science Monthly Volume 28

1886. A Thinking Machine, Gentleman's Magazine, Vol. CCLX

1886. Fish Out of Water, in Popular Science Monthly Volume 28

1886. A Thinking Machine, in Popular Science Monthly Volume 28

1886. Thistles, in Popular Science Monthly Volume 30, November 1886

1887. A Mount Washington Sandwort, in Popular Science Monthly Volume 30

1887. Among the Thousand Islands, in Popular Science Monthly Volume 31

1887. The Progress of Science from 1836 to 1886, in Popular Science Monthly Volume 31

1887. American Cinque-Foils, in Popular Science Monthly Volume 32

1888. Gourds and Bottles, in Popular Science Monthly Volume 33

1888. A Living Mystery, in Popular Science Monthly Volume 33

1888. Evolving the Camel, in Popular Science Monthly Volume 34

1889. From Africa, Gentleman's Magazine, Vol. CCLXVII

1889. Genius and Talent, in Popular Science Monthly Volume 34

1889. Plain Words on the Woman Question, in Popular Science Monthly Volume 36

1890. The Girl of the Future, Universal Review, Vol. VII.

1891. Democracy and Diamonds, The Contemporary Review, Vol. LIX

1892. A Desert Fruit, in Popular Science Monthly Volume 41

1893. Ghost Worship and Tree Worship I, in Popular Science Monthly Volume 42

1893. Ghost Worship and Tree Worship II, in Popular Science Monthly Volume 42

1897. Spencer and Darwin, in Popular Science Monthly Volume 50

1898. The Romance of Race, in Popular Science Monthly Volume 53

1898. The Season of the Year, in Popular Science Monthly Volume 54

Poetry

Grant Allen wrote various poems published in many magazines etc. We have not listed them here but hope to record some of them in the future.

www.ingramcontent.com/pod-product-compliance
Lightning Source LLC
Chambersburg PA
CBHW060626210326
41520CB00010B/1486